SPACED OUT!

An Extreme Reader . . .
from Warps
and Wormholes
to Killer Asteroids

BY BILL SCHELLER

Dedication

To David, with hopes that he'll remember to call home regularly if he's ever on another planet.
— WGS

Acknowledgment

A very appreciative "thank you" to Michael Cirone.

AND NOW
A MESSAGE FROM OUR CORPORATE LAWYER:

"Neither the Publisher nor the Author shall be liable for any damage that may be caused or sustained as a result of conducting any of the activities in this book without specifically following instructions, conducting the activities without proper supervision, or ignoring the cautions contained in the book."

THE PLANET DEXTER
GUARANTEE

If for any reason you are not satisfied with this book, please send a simple note telling us why (how else will we be able to make our future books better?!) along with the book to

The Editors of Planet Dexter
One Jacob Way
Reading, MA 01867-3999

We'll read your note carefully, and send back to you a free copy of another Planet Dexter book. And we'll keep doing that until we find the perfect Planet Dexter book for you.

Cool, eh?

A Hitchhiker's Guide to Spaced Out!

WHERE DOES SPACE BEGIN?

This is a book about "space," but just what *is* "space"? For example, are you walking around with your head in the clouds? Are you "spaced out?" In other words, just how far into space have you already gone?

This isn't a trick question. It all has to do with the bigger question of just where space begins. In one sense, you might figure that space starts off where solid ground ends. That would mean that your feet are on the ground, and the rest of you is in space.

It isn't that simple—not as long as you're short enough to be breathing air (if not, bend down, take a deep breath, and go sign a big-money contract with the NBA or WNBA). Before you actually get into space, you have to pass through the Earth's *atmosphere*, which is as much a part of Earth as soil, rock, and water.

Think of it this way: our planet is made up of matter in its three basic forms—solid, liquid, and gas. The atmosphere is the gas part of the Earth, and it begins where the other two leave off. The atmosphere contains the air that we breathe and some of the water that is essential to all life on Earth. The atmosphere also protects us from the Sun's ultraviolet rays and all but the largest of incoming meteors. The atmosphere holds in the Sun's heat, which is why it's warm enough to live on Earth but not on the Moon, which has no atmosphere. The atmosphere is also where weather happens.

The atmosphere is a gas— and we need it.

Other than space, what we on Planet Dexter most like, is to hear from readers. If you think that in this book we should have written about a subject but didn't, or if you've been to space or have had visitors from space, or if you like or don't like our book, we'd love to hear from you. Seriously! Please zap us via the Internet at pdexter@awl.com or write to us at:

THE EDITORS OF PLANET DEXTER
ONE JACOB WAY
READING, MA 01867-3999

And be sure to visit us at:
www.planetdexter.com

7

We live on the bottom of the atmosphere, like crabs live on the bottom of the sea. This is where the oxygen we need is densest, and it's where the weather is. And since gravity holds the atmosphere against the solid part of the Earth, it's also where the air pressure is heaviest—and best for humans and other living things.

As you head farther out, you enter the *stratosphere*—which is anywhere between about 9.5 and 31 miles above the dirt in your backyard. The temperature is pretty consistent and for the most part, there's no weather (the odd high cloud or large thunderstorm sometimes pokes its way into the stratosphere). Out there, gravity's less in charge (gases released during volcanic eruptions will remain here for years before settling down toward earth) and the air is thinner and has less oxygen. That's why jet planes, which fly in the stratosphere, have pressurized cabins; and why Himalayan mountain climbers, who get almost but not quite up to the stratosphere (the tip of Mount Everest is about 8.8 miles high), usually carry tanks of oxygen.

Head up even higher, through the layers of atmosphere called the *mesosphere* and *thermosphere*, and the air gets even thinner. There's also a lot less oxygen—bring your balloons, because lighter gases such as helium and hydrogen take over here.

Finally you reach the *exosphere*, where the air is thinnest of all. There's no lid on top of the exosphere, no place exactly where you can say, "All right, this is it"—but somewhere between 600 and 700 miles up the soup* pretty much runs out.

Now, you're in space.

*The air that makes up the atmosphere is a soup made of nine basic elements. Two of them, nitrogen and oxygen, account for 99 percent of the whole thing — roughly 78 percent nitrogen, and 21 percent oxygen. The other gaseous elements are the spice in the soup — stuff like carbon, joined with oxygen to make carbon dioxide (handy for putting the fizz in soda) and the ever-popular neon and helium, hot items if you want to light up signs or fill balloons.

AND THEN WHAT?

There's space, and then there's space. When astronomers or space scientists talk about *interplanetary* space, they mean the region within our own Solar System—the Sun, the nine planets that orbit around it, and their moons, along with the asteroid belt and the comets that follow a loopy path around the Sun (more on these later).

Galactic space means everything within the Milky Way galaxy, which is the enormous cluster of stars our Solar System is part of, and which you can see as a long splash of milky whiteness on perfectly clear nights when you're far away from city lights.

Finally, there's *intergalactic* space. Are you following the pattern? You got it—this is the whole Universe, the Milky Way and all the other galaxies, and all the space between.

Fancy Word Feature
Impress Friends and Family

troposphere: The lowest part of the atmosphere, starting at the surface of the Earth.

atmosphere: The gaseous mass encompassing the Earth, held by gravitational force.

stratosphere: A part of the atmosphere below the mesosphere, where the temperature is pretty consistent.

mesosphere: That part of the atmosphere, from 20 to 50 miles above the Earth, marked by a temperature range that drops from 50°F to −130°F with increasing altitude.

thermosphere: What sits between mesosphere and exosphere. Temperatures here increase the higher one goes because the thermosphere's air is not very dense, causing it to be greatly affected by the sun's radiation.

exosphere: The outermost portion of the atmosphere, beginning at about 300 miles above Earth.

napiform: Shaped like a turnip.

Urban legends. If a story sounds true, but also seems too good to be true, it's probably an *urban legend.* At one time or another, just about everyone's heard of the poodle that exploded when its owner tried to dry it in a microwave . . . or of the person who brought home a strange-looking Chihuahua puppy from Mexico, only to learn it was really a rat . . . or of alligators roaming about the sewers of New York City.

Experts in urban legends (what a job, eh?) claim that a good urban legend should:

1. contain a *grain* of truth that implies that the *entire* story is true,

2. reflect contemporary fears,

3. imply that the person telling the story knows the person who knows the person who witnessed or is involved in the story, and

4. it sure helps if it shows up in major media (like *The New York Times*, *The Wall Street Journal*, or a really weird Planet Dexter book), either as rumor or fact.

If You Think You're Having a Bad Day

(An Urban Almost in Space Legend)

Fire authorities in California found a corpse in a burnt-out section of forest while assessing the damage done by a forest fire. The deceased male was dressed in a full wetsuit, complete with a dive tank, flippers, and face mask.

A postmortem examination revealed that the person died not from burns but from massive internal injuries. Dental records provided positive identification.

Investigators set about determining how a fully clad diver ended up in the middle of a forest fire. It was revealed that, on the day of the fire, the person went for a diving trip off the coast—some 20 miles away from the forest. The firefighters, seeking to control the fire as quickly as possible, called in a fleet of helicopters with very large buckets. The buckets were dropped into the ocean for rapid filling, then flown to the forest fire and emptied.

You guessed it. One minute our diver was making like Flipper in the Pacific, the next he was doing breaststroke in a fire bucket 500 feet in the air. And then—BLAM!—he extinguished exactly five feet, ten inches of the fire.

THE SUN

If you live on Earth, the single most commonly visible object in the sky is the Sun. Even if it isn't doing what we call "shining," it *is* shining. You can see the proof of this all around you, mainly in the fact that you aren't dead. Solar energy is the basic engine of life on Earth, and solar gravity is what keeps us where we can get just the right amount of it.

The Sun is a star, and that means it is a nuclear furnace fueled by hydrogen. The process that keeps the Sun going is called *nuclear fusion*. In the simplest terms, what happens in nuclear fusion is that atoms of hydrogen collide under great pressure, creating helium atoms—and also releasing a tremendous amount of energy. It's what happens when a hydrogen bomb explodes. (An atomic bomb, on the other hand, releases its energy when uranium atoms *split*. That's called *nuclear fission*.)

If the Sun is a big hydrogen bomb, why doesn't it just explode and get it over with? The answer has to do with gravity. The gas cloud that drew together to become the Sun—or any star—gained in gravity as it increased in mass, reaching the point where gravity actually contains its powerful fusion reaction. The Sun is a controlled reaction, a constantly exploding bomb that holds itself together. It uses up hydrogen at an incredible rate—10 billion pounds per second or more—but it has so much of this fuel that it's been burning for 5 billion years, and will burn for 5 billion more.

So, you've got time to finish this book.

Dear Planet Dexter. Most books now say our sun is a star. But it still knows how to change back into a sun in the daytime . . . .

15

THE MOON

Let's start close to home—and in space terms, the Moon is very, very close. Its distance from Earth varies from roughly 226,000 to 252,000 miles, making it by far the nearest object in space and the only one human beings have ever visited. The Moon orbits around Earth—in other words, it's the original Earth satellite.

The Moon is a little bit less than one-quarter the size of Earth, measuring roughly 2,160 miles in diameter (Earth's diameter is about 8,000 miles). It takes the Moon 27 days and just under 8 hours to make the full-circle trip around Earth, but as it revolves it behaves differently from the way Earth does as it travels around the Sun. The Moon doesn't spin on its axis, which is the imaginary line running through the north and south poles of a planet or moon. So the same side faces Earth all the time, year in and year out, while the other side—the "dark side" of the Moon—is always out of our view. We didn't get a look at it until the first time we sent spacecraft into lunar (Moon) orbit.

It was the Moon that gave humankind its idea of months (notice the connection between the words?). Because of the way sunlight is reflected off the Moon's surface—it has no light of its own—it looks as if it is changing its shape, from full to crescent to half to crescent again and finally to nothingness, over the course of a 29 1/2-day cycle.

FULL   HALF   CRESCENT

This slow-motion disappearing act is due to changes in the positions of the Sun, Earth, and Moon in relation to each other, but primitive people didn't know that. To them, it was a handy clock, and so we have months of roughly 30 days.

Aside from operating as a clock, the Moon also does the important job of changing the tides in Earth's oceans. The tides rise because of the effect of the Moon's gravity as Earth turns on its axis. Without this continual washing back and forth of sea water on the shores and in salt marshes, many basic and important forms of life could not survive. In fact, a few scientists have suggested that without the Moon sloshing the tides around, life on Earth might not have gotten started at all.

Years of looking through telescopes, and finally visiting the Moon in person, have shown us that our little satellite has a really bad complexion. Why all the craters? Actually, Earth would look the same way if we didn't have our atmosphere. Small meteoroids burn up when they smack into our layer of air, and the craters made by bigger meteors have mostly been eroded away by wind and water. The Moon has no such protection or activity—its gravity may yank our oceans back and forth, but it isn't strong enough to hold an atmosphere. This is because gravity increases with size, and the Moon is just too small. A meteorite hits, a crater forms in the lunar rock and dust, and there you are—it's still around, looking brand-new, a million years later.

How did we wind up with the Moon, anyway? There are several theories. (One thing about science, there's never just one theory.) Some scientists say that Earth and its Moon were formed together. Others say that back when the Solar System was young, and everything was pretty much made of hot glop, the Moon tore off from Earth (after all, the two bodies share many of the same basic materials) and went into orbit around its parent, which had the stronger gravity.

There's also a theory that Earth's gravity captured the Moon, which had been out there on its own. And yet another theory proposes that the Earth-Moon team got started when a large object struck the Earth, and then bounced back into space bigger than it was when it hit. Roll up two snowballs (or mudballs if there's no snow handy), a big one and a little one, then press them together and pull them apart to see if you can duplicate this event.

THE NEARER PLANETS

Of the nine planets held in orbit by the Sun's gravity, the first four are pretty much alike in that they are small and made of rock. Earth, the third planet out from the Sun, fits in with this crowd even though its special conditions—namely, water and life—set it apart. Mercury is closest to the Sun, then comes Venus. The next orbit after Earth's belongs to Mars, which has two moons of its own.

Little Mercury is in the hot seat. The smallest of all the inner planets—it has a diameter of about 3,000 miles, a little more than a third of Earth's—Mercury orbits the sun at a distance that varies from 29 to 43 million miles. This means that the Sun is not at the exact center of Mercury's orbit, making this an "eccentric" orbit. Fortunately, the Earth does not have this kind of orbit. If it did, it would be too hot for us to survive during the part of the year when we were close to the Sun—even if we glopped on a really good sunblock.

By the way, that's just what a year is—the time it takes for a planet to make a complete trip around the Sun. For us, it's about 365 days. For little Mercury, it's only 88 Earth days. That's how the planet got its name. Mercury was the super-fast messenger of the gods in Roman mythology, which is why you always see pictures of him with wings on his feet.

Like the Moon, Mercury isn't big enough to have gravity capable of holding down an atmosphere. Also like the Moon, Mercury is pockmarked with craters from the impact of meteorites that met no resistance on their way to the surface. Another thing that happens when a planet doesn't have an atmosphere is that there's no way of holding in heat. An atmosphere serves as insulation, like fiberglass in a house or a winter parka on your body.

So, even though Mercury is the closest planet to the Sun and its surface temperature on its daylight side is probably around 800° F, the the side facing away from the Sun—the nighttime side—might be 300° F *below zero*. It gets to stay that way for a long time, since Mercury turns so slowly on its axis that daytime lasts for about 176 Earth days.

Venus is another story. Venus is about the same size as Earth, and it has an atmosphere. Boy, does it have an atmosphere—unmanned Russian and American probes have discovered that "air" on Venus consists almost entirely of carbon dioxide. Reflection of sunlight off this thick blanket is what makes Venus the brightest planet in Earth's night sky. The lovely appearance of Venus, especially at dawn and twilight during certain times of the year, caused the ancients to name the planet after the Roman goddess of beauty.

But the thick, cloudy atmosphere of Venus also makes it the hottest planet—temperatures probably reach 900° F on the surface, day and night. That carbon dioxide blanket insulates Venus so well that solar heat just doesn't escape, the way it does from Mercury, even though Venus is 67 million miles from the Sun, a lot farther away than Mercury is. When environment-watchers here on Earth talk about a *greenhouse effect* from too much carbon dioxide in our atmosphere—the result of burning too much fuel that releases this gas—they can point to Venus as an extreme example of this situation. (The *greenhouse* part of that term, by the way, comes from the fact that glass traps heat in greenhouses which are used for growing plants.)

Venus's atmosphere is also extremely heavy. If you were on the planet's surface, it would feel 100 times as heavy as the air overhead on Earth. Of course, you wouldn't feel it at all, because you would be cooked like a lobster and squashed like a bug.

From pictures taken by space probes that have penetrated Venus's atmosphere, and even by Russian craft that have actually landed on the planet, we've learned that the surface has mountains, volcanoes, craters, and dusty plains. Weather—not water, but wind—has blown sand dunes around. It's all a lot different than the steamy venusian jungles some people used to imagine many years ago, when they thought that dense, cloudy atmosphere was full of water vapor. And it certainly doesn't look like anyplace you'd name after the goddess of beauty.

Skipping over Earth, which of course you know everything about already, we come to the outer-most of the small, rocky planets—Mars. Mars also has an eccentric orbit, which causes it to vary from 35 million to 63 million miles in distance from Earth.

Mars was the Roman god of war. Since the planet Mars appears to have a reddish tint, the ancients looked at it and thought of blood, and gave the Red Planet its name. Actually, Mars *is* red—the same rusty color as much of Earth's desert lands. For Mars is a desert, a world where water may once have existed but is long gone (except for a small amount that may be locked up in what seem to be icecaps at the planet's north and south poles).

One of the things that has gotten people thinking that Mars might support life is the fact that these icecaps seem to shrink and expand as the martian seasons change. (Yes, Mars does have seasons, because it is tilted on its axis the same way Earth is. This means that, depending on the time of year, either the north or south pole is tilted toward the Sun.) From Earth, it even looks as though there are green areas that expand as the icecaps shrink.

But close-up views of Mars taken by spacecraft that have flown by or even landed on Mars show that these green areas are not fields of vegetation. They might be covered with some mineral that changes color with the martian seasons, or they might just look green because they're next to red areas. Our eyes play color tricks on us. When a large patch of gray borders on a large patch of red, the gray area can look green.

Anyway, that shrinking and expanding of the "icecaps" on Mars is not caused by water melting and re-freezing. Mars is too cold for water ever to melt. The changes are probably caused by carbon dioxide changing back and forth from solid to liquid. Carbon dioxide also seems to be the main ingredient in the martian atmosphere, which is only about 1 or 2 percent as dense as ours. (Mars is about halfway between Earth and the Moon in size, so it is a lot weaker than Earth in the gravity department).

Was there ever water on Mars—or, at least more water than the tiny amount that might be locked in the icecaps today? It looks that way, since along with its barren, rocky deserts and huge volcanoes, Mars has gullies and canyons that look like dry riverbeds. If it was water that eroded these features, then we have to figure that Mars was warmer at one time.

With the surface of Mars looking as bleak as it does, the chances of Marvin or any other Martians turning up seem to be about zip. But in 1996, scientists found something that might show that there was some kind of life on Mars, at a long-ago point in its history. They found this evidence on Earth, of all places.

The object the scientists are studying was found in Antarctica. It's a piece of rock that was once part of Mars (we can tell this by what it's made of), and was knocked loose from the Red Planet by a meteorite and later smacked into Earth. Inside the rock are what look like the fossils of tiny organisms, sort of like Earth bacteria. These little dead creatures, if that's what they are, are really, really dead—several billion years worth. So, even if researchers do decide that this proves Mars once had life, it may have become extinct long ago as the planet cooled and dried up.

Dexterdrome!

(Dexterdromes — some folks call them palindromes — are phrases that are spelled the same way backward as forward.)

"Do orbits all last?" I brood.

Weird!

Astronaut Buzz Aldrin was the second man to step onto the moon. His mother's maiden name was *Moon*.

STAR WARS FACTS

In July 1973, George Lucas first approached Universal Studios to see if there was interest in his idea for a film idea he called *Star Wars*. Universal turned him down (bad move).

Star Wars opened about four years later, on May 25, 1977. Within three months it had grossed $100 million, shattering all film revenue records up to that time. Within just six years, *Star Wars* tickets sales, worldwide, were over half a *billion* ($524 million).

Question: Where did the idea for Chewbacca come from?

Answer: Lucas got the idea for Chewbacca one day as he watched his wife drive off with their Alaskan malamute, Indiana (who would later inspire the leading character's name in *Raiders of the Lost Ark*). Lucas liked the way the large shaggy mutt looked in the passenger seat. So he decided to create a character in the film that was a cross between Indiana, a bear, and a monkey.

Q: On the first screening of the film for executives of 20th Century Fox (who hated it—some sleeping through it, and others not getting it), what change was suggested for C-3PO?

A: That a moving mouth be added to the robot; otherwise, moviegoers could not possibly understand how he could talk.

Q: On the film's videotape version, why do Aunt Beru's lips appear not to be moving with her speech?

A: Because her lines are all dubbed over. After the original theatrical release, Lucas felt that the real Beru's voice was too low, and needed to be changed.

Oops, a blooper.

Right after the Death Star is destroyed, as Luke is descending the ladder out of his X-Wing and Leia comes running up to him, Luke mistakenly calls out her real name, yelling "Carrie!" (Princess Leia is played by the actress Carrie Fisher).

Oops! Duck!

Luke, Leia, Han, and Chewie are inside the Death Star's trash compactor. Suddenly it begins to compact. Luke yells into the comlink and the camera cuts to C-3PO's comlink sitting on the table. Just then a security door opens, stormtroopers march in, and—blam!—the actor playing the trooper behind and on the right of the lead trooper accidentally hits his head on the door frame.

CANALS ON MARS

In 1877, Italian astronomer Giovanni Schiaparelli saw what looked to him like straight lines running across the surface of Mars. He named them *canali*, which is Italian for *channels*. Schiaparelli didn't say how he thought they got there. To him they were just straight lines, and *canali* was as good a word as any.

Along came Percival Lowell, an American astronomer. He translated Schiaparelli's *canali* as *canals*—which is another meaning for the Italian word—and took a look for himself. Yup, he decided, they're canals. Lowell went on to cook up an explanation: the canals were built by a race of intelligent beings to carry water from the planet's polar ice-caps, at a time when Mars was drying up. The scheme worked for a while, he wrote, but eventually the water ran so low that the Martians died out.

Nobody believes Lowell's theory anymore. In fact, modern close-up photos of Mars don't show a system of canals at all. Now, if Schapiarelli had said that he had seen *cannoli*, at least we could go to Mars for dessert.

Fancy Word Feature
Impress Friends and Family

cannoli: A fried pastry roll with a creamy, usually sweet filling.

seleneologist: Somebody who studies the moon.

COMETS, ASTEROIDS, METEORS, AND METEORITES

Figuring out the Solar System—and the whole universe, for that matter—is like putting together a clock and constantly finding parts that don't fit. It's nice to think in simple terms of the planets revolving around the Sun, and the planets' moons revolving around the planets. That is the way things work, but then there are those leftover parts . . . .

Comets **are one such complication.** A comet has been described as a dirty snowball, a mass of frozen materials such as carbon dioxide and ammonia mixed with dust. The head of a comet might be only a few miles across at its core, while its tail can be tens of millions of miles long. The tail is caused by partial evaporation of the snowball as the comet approaches the Sun.

Comets orbit the Sun the way the planets do, but in paths that look nothing like planetary orbits. The planets all spin around the Sun in the same direction, and their orbits are pretty much elliptical. But comets are a bunch of oddballs. Some orbit the Sun one way, some the other. And the shape of their orbit is a flattened circle, an ellipse, so that they sail way out of the Solar System's bounds before looping back around the Sun. Some orbit in a short ellipse and show up fairly often— like Halley's comet which comes back every 75 years. Others turn up only every few thousand years. Comet Hale-Bopp, which many of us saw when it zipped by during the winter of 1995-96, is one of these.

Asteroids **hang out a little bit closer to home.** Like a planet that can't pull itself together (which many scientists think is exactly what they are), they orbit the sun in a belt that lies between Mars and Jupiter. There are probably more than 100,000 of them, and about 1,500 are big enough to have names. Vesta, the only one you can see without a telescope, is about 240 miles across. They are leftovers from the beginning of the Solar System, the homeless motorcycle gang of the heavens.

What about *meteors*? If we want to get picky, the word meteor doesn't describe an object at all, but the streak of light we call a shooting star when we see it flash across the night sky. These flashes come in packs, called *meteor showers*, at certain times of the year. The particles that cause the flashes and the showers are called *meteoroids*, and most of them are about the size of a grain of sand. When they hit Earth's atmosphere they burn up from the friction, and that's where the flash comes from.

Astronomers believe most of the particles that make up meteor showers are what amount to comet dandruff, fragments that trail off from comets when they boomerang around the Sun. Hundreds of millions of them a day hit Earth, and the dust they leave behind—each day—amounts to several thousand tons, probably more dust than there is in your entire room.

Then there are *meteorites.* These are the space rocks big enough to make it through the atmosphere, smack into the Earth's surface, and occasionally get found. They don't fall during meteor showers, and they aren't comet dandruff. They probably formed in much the same way as the asteroids, only they don't have the manners to stay out past Mars with their hoodlum friends. There's only one recorded occasion when a person was hit by one—only a minor injury—but there are some awfully big craters that show where huge meteorites have landed in the past. One crater, in Arizona, is 4,000 feet across and 600 feet deep. The meteorite that dug that hole was not out to cause *minor* injuries.

The only known
meteorite fatality
happened when a dog
was hit in Egypt
in 1922.

WHAT ARE THE ODDS OF GETTING FLATTENED?

And What About the Poor Dinosaurs?

The bad news is that any meteorite bigger than a truck would flatten a city with the explosion of its impact, and a meteorite a few miles across could wipe out most life on Earth. A lot of scientists think this is what happened to the dinosaurs—who were doing just fine, thank you—until about 65 million years ago when they suddenly died out.

Being mammals, we always like to think that we just somehow outsmarted dinosaurs, or tickled them to death with our fur. But it was probably a meteorite that did them in. What may have happened was that when the meteorite struck Earth, it raised such a cloud of dust—and smoke from forest fires it set— that sunlight couldn't get through the atmosphere for years, causing most life—plants and animals—to die out.

The good news is that this kind of thing is likely to happen only once every few million years. Just in case, astronomers are trying to make a list of the asteroids, meteoroids, and comets that might be heading down our part of the sidewalk. (A comet would be no fun to get hit by either.) People are even talking about how to win an argument with one of these clunkers, either by destroying or deflecting it with a bomb, or by landing a rocket on it and using thrusters to change its course. If this sounds like a big video game, well, where do you think the game people get their ideas?

Fancy Word Feature
Impress Friends and Family

herbivores: Plant-eating animals.

carnivores: Flesh-eating animals.

taurophobia: Fear of bulls.

THE BIG GASBAGS

Things get really different once you get past Mars and the asteroid belt. First of all, the distances between planets, and also between each planet and the Sun, get much greater. It's as if Mercury, Earth, Venus, and Mars are huddling close to the campfire, while the other guys are way off in the bushes. Second, the outer planets—except for tiny Pluto, which is in a class of its own—are a *lot* bigger than us guys by the fire.

Jupiter, Saturn, Uranus, and Neptune are known together as the Giant Planets, but we might also call them the Gas Planets. They are all far larger than Earth—Jupiter, the largest (Jupiter was King of the Roman gods) is 1,300 times as big as Earth, if we're talking about the space it takes up. But for all its size, Jupiter is a lightweight. It's probably only about 300 times as heavy as our home planet. The other three, out beyond Jupiter, are also big puffballs.

Why are the big planets so light? It has to do with the strange stuff they're made out of. If you were to start at the top of Jupiter's atmosphere, you'd go down through several hundred miles of ammonia ice clouds. Ammonia is this horrible-smelling stuff people used to use a lot for cleaning—so these would not be very pleasant clouds. Aside from the ammonia, Jupiter's atmosphere is largely methane. Methane is a mixture of carbon and hydrogen; on Earth it rises from rotting vegetation and cow poop. If somebody puts cow poop in a container and lets it rot, the methane gas that comes from it can then be lit with a match.

(DO NOT try this at home. Never play with matches, or with cow poop.)

Down below, it would be hard to tell where Jupiter's atmosphere ends and its surface begins. The atmosphere—which has fun features such as 4,000-mile-per-hour winds and a 300-year-old hurricane called the *Red Spot*—probably turns to slush down near the "ground," which itself may be an ocean of liquid hydrogen. As far as anybody can guess, the hydrogen gets denser and denser toward Jupiter's center, which might be a core of rock about the size of the Earth.

If we ever land anywhere near Jupiter, it would have to be on one of the planet's 15 moons. All of these outer gas planets collect moons like baseball cards. Saturn has 21 or more, Uranus has at least 15, and Neptune has 8. Saturn, of course, also has its famous rings, which can be seen only through a telescope. As near as we can tell, the rings are made of ice crystals. They're 28,000 miles across from outside to inside, but only about 10 miles thick. They may be the remains of a torn-apart ice planet.

Saturn and Uranus (both also named after Roman gods) and Neptune (named for the Roman sea god—it's a bluish planet) all seem to be like Jupiter in construction, with poisonous (to us) atmospheres that thicken near the surface until they blend in with the soupy surface. Then, in the center, there's a rocky core.

The gas planets are just the opposite of a Tootsie Roll pop.

PLUTO:
THE ODD DOG OUT

Pluto is the odd one out. It's practically a pebble, smaller even than our Moon—in fact, some scientists think it's a moon of Neptune's that managed to escape the bigger planet's gravity. A rock surrounded by frozen methane, so cold and far away it seems like something the Solar System forgot, Pluto wasn't even discovered until 1930. It chugs along out there, with one little moon of its own, taking almost 248 Earth years for each trip around the Sun. It has an odd orbit—sometimes it ducks in closer to the Sun than even Neptune. So Pluto's the farthest-out planet *most* of the time . . . but not *all* the time.

NO, BAD DOG!

BEYOND THE SOLAR SYSTEM

That's it for our immediate family—the system of planets and smaller objects that orbit around the Sun. But the Sun is part of a far larger family of stars, the Milky Way galaxy.

Even without telescopes, we've always been able to see the Milky Way by looking into a clear night sky at its billions of stars, all seeming to be packed together so closely (they aren't) that they no longer look like individual points of light. But until this century, we had no way of knowing that what we were looking at happened to be merely our own home galaxy, and that there were billions more of these vast families of stars scattered throughout the Universe and separated by empty gulfs of space.

Using the small, simple telescopes they had available, astronomers as far back as 200 years ago noticed fuzzy shapes in the night sky that weren't stars, but didn't look like anything else familiar, either. They called these shapes *nebulae (nebula* is the singular form; it means *cloud* in Latin, which they should teach you in school), and figured they were clouds of dust and gas.

Some of them *are* clouds of dust and gas, and they're located right here in the Milky Way. But even back in the 1800s, some star watchers were wondering if some of the nebulae might actually be clusters of stars, far from our own. Proof of this finally came in the 1920s, when the great astronomer Edwin Hubble went to work at the huge new 100-inch (the lens diameter) telescope at Mount Wilson, California. Hubble taught the world about galaxies—the 100 billion or so star systems in the Universe, of which the Milky Way is only one. He did it by identifying stars within those nebulae that were much closer to each other than they were to us. Because of his discoveries, astronomers named the Hubble Space Telescope after him.

The Milky Way is a spiral galaxy, like most of the others. When we take photographs of the others, they look like dazzling white pinwheels—and, like pinwheels, the galaxies are actually spinning. (We can't take a photograph of our own galaxy, the Milky Way, because we're in it about two-thirds of the way out on one of the arms of the spiral.) Our Sun and its planets make the full circle every 250 million years. In all its history, the human race has been along for only a small part of this ride—about 3 degrees of the circle—even if you go way back and include some humans that you probably wouldn't want to hang out with.

What's in between the galaxies? Not a whole lot that we can understand—and maybe a lot more that we can't. On the one hand, the spaces between galaxies seem as empty as the spaces between stars. But space isn't really empty. It's a region of gas and dust, spread out so thinly that compared with our atmosphere, it's practically a vacuum. Mostly, the gas molecules out there are hydrogen. Hydrogen is the most common element in the Universe.

In recent years, though, scientists studying the history and evolution of the Universe have been wondering if their theories have to depend on some other form of matter between the stars, something that doesn't shine and that we can't see. They call this stuff *dark matter*. If dark matter does exist, no one is quite sure just what sort of particles it's made of.

THINKING IN LIGHT-YEARS

As long as we're talking about our Sun and its family of planets, miles or kilometers will do. But when it comes to the stars—even the closest ones—the numbers we have to attach to those old-fashioned measurements get so big that they start to fall off the page.

For example: Do you know how far across our Milky Way galaxy is? Take a deep breath, and start counting "one, two, three. . . ." The Milky Way is 600,000,000,000,000,000 miles across, maybe a couple of miles more if you take Route 80 around Paterson, New Jersey, instead of heading straight through downtown. That's *six hundred quadrillion miles*. It would get ridiculous writing numbers like that. So astronomers came up with the concept of the *light-year*. One light-year is the distance that light can travel in one year.

Light doesn't seem to travel at all, does it? You turn on the switch, and there it is. You look at something, and you know you're seeing it just as it is this instant, because the light traveling from it (which is, after all, what you're seeing) gets to you immediately.

Actually, it's not quite immediately. Light does have a speed, and that speed is 186,000 miles a second. That adds up to 6,000,000,000,000 (6 trillion) miles in a year. Here on Earth, that means almost nothing—the whole globe is only about 25,000 miles around. Even moonlight reaches us in just a fraction more than a second. Sunlight? That gets to Earth in a little over 8 minutes.

But once we get out into deep space, where we're figuring the distances between stars and galaxies, light starts taking some serious time to get from one place to another. Even the next nearest star, Alpha Centuari, is 4.3 light-years away. The North Star, which has been guiding sailors for thousands of years, is 680 light-years away.

The Milky Way is 100,000 light-years across—an easier number to remember than that string of zeros we wrote down earlier, but still enormous. The Andromeda galaxy is 2 million light-years distant from the Milky Way, and it's 900,000 light-years across. Some galaxies are only a tenth the diameter of the Milky Way; others are a hundred times wider.

As to how far these galaxies keep on going—and whether it's just galaxy after galaxy, forever—that's a question to tackle when we talk about how the Universe might have formed, and what is going to become of it. We'll get to that—but first, let's get back to Earth, and look at the stars not the way Edwin Hubble did, but the way our ancestors did, a few thousand years back.

THE CONSTELLATIONS

In ancient times, the sky seemed a lot darker at night than it is now. The sky was actually no different, of course—it's just that there was none of what we call *light pollution*, which is the glow that rises up from our cities and suburbs and makes it impossible to see all but the brightest stars. Two thousand years ago, you could look up at the night sky even in a city and see stars the way you can see them now only if you're out in the middle of nowhere.

Now, combine this fact with the fact that people in those days—and we're talking mainly about the Greeks and Romans—had the kind of imaginations that allowed them to build myths around the things they saw in the natural world. Throw in the spare time you get from not being able to watch TV, and you'll start getting an idea of how the constellations were named and described.

The constellations are pictures in the sky, patterns of stars that suggested people, animals, and objects to the ancients (other societies saw them too, but our constellation names come down to us from the old civilizations—Greek, Roman, and so on—which hung out around the Mediterranean Sea). You may think that it takes quite a stretch of the imagination to come up with some of these things, but again, remember: No TV.

Another thing to keep in mind about the constellations is that just because a group of stars as seen from Earth make up a picture, that doesn't mean the stars are anywhere near each other. For example, the fainter stars are not necessarily smaller; instead they may be much, much, much, much, much farther away. They exist in three dimensions, but we read them in two.

Imagine kids sitting around in Athens 2,500 years ago—

"GET TO BED!"

"Aw, Ma, can we just stay up and look at a few more constellations?"

47

WHAT'S YOUR SIGN?

You will probably recognize the names of some of the constellations, even if you're not a star-watcher. If you can't remember where you saw the names before, take a look at the horoscope page in your newspaper.

That's right. Twelve of the constellations have the same names as the signs of the Zodiac, a circular symbol first developed by the Babylonians about 2,500 years ago. According to the ancient system of belief called *astrology*, each of these signs of the Zodiac is associated with certain aspects of the human character.

Astrology was practiced by people all over the world, including people in societies that had no contact with each other—the Mayans, Hindus, Chinese, Babylonians, Hebrews, Greeks and

Romans, and medieval Europeans. It survives in its simplest form in those columns in the newspapers, though there are many people today who take it a lot more seriously than that. Astrology is a very complicated subject—it isn't just a matter of saying Scorpios are this way, and Leos are that way. But when all is said and done, it comes down to one thing: Do you want to believe that any part of your character or your fate has to do with what position the stars were in when you were born? Do you want to make decisions based on where the stars are?

Suit yourself. We're flipping a coin.

THE CHEAPO PLANETARIUM

Places like the Hayden Planetarium, at the Museum of Natural History in New York City, project fantastically accurate maps of the heavens onto the inside of big domes by using optical instruments that cost millions of dollars. But you can project constellations onto the wall of your room for free, using a flashlight and a 10- or 12-ounce paper cup.

Here's how:

1. Drink something cool from the cup.

2. Copy one of the constellations (see next two pages) onto the bottom of the cup.

3. GRAB AN ADULT! (Having adults help you out in situations like this is why they exist, really!)

$3\frac{1}{2}$. Place the cup open-side down and instruct (this is known as "taking a leadership position") your handy adult to use a pen to punch a small hole in the cup's bottom for each star in the constellation.

4. Unload adult.

5. In a dark room (BOO!), put a lit flashlight into the cup and the stars of the constellation will appear on the wall or ceiling!

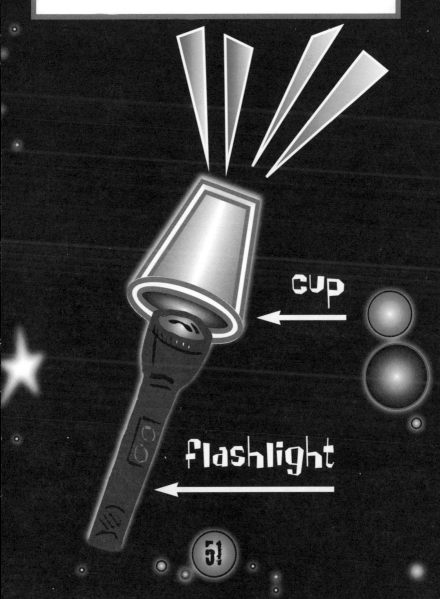

cup

flashlight

CONSTELLATIONS

 Aquarius

Ursa Minor

Canis Major

Cassiopeia

Cygnus

Gemini

Leo

Orion

Sagittarus

Scorpius

Dexter Major

53

STARS COMING AND GOING

Identifying and naming all the stuff in the Universe, and then figuring out how it moves, took up the first few thousand years of skywatching. Over the past century, though, scientists have begun to think more seriously about what the Universe is made of, how it came into existence, how old it is, and what's going to become of it. This science is called *cosmology*, and it has led us into some pretty strange corners. No explanation ever offered by primitive mythology for how the Universe got here was weirder than what the cosmologists have learned so far.

As we saw when we took a look at our own Sun, stars are born and they die. The Sun, which is pretty much an ordinary, medium-sized star, has been around for 5 billion years and has enough hydrogen fuel to keep going for 5 billion more. After that, like other stars its size, it will expand outward and then collapse. First it will be a hot, dense star-ruin that astronomers label a *white dwarf*, and billions of years later still it will become a cool ember called a *black dwarf*.

Bigger stars, as they run out of fuel, will explode

with the firecracker dazzle that we call a *supernova* on the rare occasions when we come across one while scanning the skies. And some of those blown-out biggies will collapse into the strange entities that no one has ever seen but that the laws of astrophysics tell us must exist: *black holes.*

A black hole is a burned-out star that has collapsed back into itself, becoming so dense that nothing—not even light—can escape it. Remember, it's mass (the bulk that causes weight), not size alone, that is responsible for gravity. An entire star collapsed to the size of an invisible pinpoint would have a gravitational pull that would—there's no other way to put it—suck, more than anything you could imagine.

But stars are born, too. The Hubble Space Telescope has even sent back beautifully clear pictures of the great gaseous clouds from which stars form. Some of the material given off by used-up, exploding stars finds its way into new stars, just as decaying plants put nutrients into the ground where new plants grow.

How do astronomers know whether a star is young or old? For that matter, how do they know anything about the temperature or distance or chemical makeup of things so far away? After all, all we can do is look at something that distant. We can't touch it, or even send up gizmos to sift through soil and analyze rocks the way we've been able to do with Mars. And we certainly can't tell the age of stars the way we date rocks and bones and cavemen's tools here on earth, using the rate of decay of carbon-14 atoms.

But looking has done the trick so far—looking through telescopes, and *listening* to the radio waves and microwave radiation given off by these far-off objects in space.

Even if you don't have a telescope, you can pick out certain telltale differences among the stars you can see on a clear night. First of all, some are a lot brighter than others—they're the ones you can see even on a not-so-clear night. On the basic scale of *star magnitude*, in use since ancient times, we give the lowest numbers to the brightest stars (they're the stars of the *first magnitude*), and higher numbers to dimmer and dimmer stars. In the city, where there's a lot of interference from lights on the ground, you might only see first-magnitude stars. Astronomers even have a way of adjusting the magnitude scale so we can tell the difference between stars that look bright only because they're close and far-away stars that really are burning brightly.

Looking at the stars with special equipment we can actually tell how old a star is, what point it is at in the course of its life, and even its surface temperature. This all became possible when we began to understand that each chemical element gives off its own wavelength of light in the spectrum. (You know the spectrum—the range of colors that light is broken into when it passes through a prism? Well, the chemical wavelength spectrum is more complicated than the light spectrum, but those rainbow colors are a useful model.) Using a device called a *spectroscope*, astronomers can tell which elements are giving off light from the surface of a star. That tells how hot it is, which in turn tells where it stands in terms of its lifespan.

Fancy Word Feature
Impress Friends and Family

cosmology: A branch of philosophy that considers the genesis, processes, and structure of the universe.

cosmonaut: A Russian astronaut.

dexternaut: A Planet Dexter astronaut.

COSMOLOGY'S BIG QUESTIONS

So here we are, tucked away in our corner of the Universe lit by billions upon billions of roaring hydrogen fusion furnaces. Some of them are cruising along in the prime of their lives, like the Sun (lucky for us). Some are nice and fresh, and others are dying. The birth and death of stars involves the release and recycling of all the chemical elements we know on Earth—hydrogen, mostly, and helium and carbon and magnesium and silicon and everything else as well. These elements form clouds of gas and dust that pull together as they build gravity; and eventually, they blow apart as their star-life ends.

It is a Universe that is doing a lot more than just sitting there.

But how did it get started? Where did everything come from? Does it have a boundary, like our

Solar System or even our galaxy? And what will eventually become of it?

These are the questions that cosmologists face. And the best answer they've come up with, so far, is the

Big Bang!

You may have heard of the *Big Bang*, the theory that the Universe began with a single explosion of matter from which everything came into being. It's one of those scientific terms that people grab on to, because it sounds simple and exciting. Well, maybe it is exciting. But it's not all that simple, because it raises as many questions as it answers.

The Big Bang. It sounds like an explosion in space, sending the stuff of galaxies and stars and planets—and us—flying out in all directions. But the cosmologists are quick to remind us that according to this theory, the Big Bang wasn't just an explosion of matter in empty space. It was an explosion that involved the creation of space itself, expanding outward where there was NOTHING—no space, no vacuum, no dark, no place, just NOTHING.

"Expanding outward" is the key term here. The reason cosmologists thought up the Big Bang in the first place is that they saw movement throughout the Universe, a movement they could identify because of something called the *red shift*. The red shift is another effect that can only be seen with a spectrograph. When a galaxy's light shifts toward the red end of the color spectrum, it's a sign that the galaxy is moving away from us. One of the things that Edwin Hubble noticed when he was studying galaxies is that they all show the red shift.

Is it our breath? Don't we shower enough? Why, of all the pinpoints in the Universe, should Earth's Milky Way galaxy be the place everything is moving away from?

But think about it. The red shift doesn't mean that everything is moving away from us, but that *everything is moving away from everything else*. The analogy that a lot of writers use has to do with raisins in raisin-bread dough. As the dough expands, all of the raisins move apart. But if you were a raisin, it would look as if all the other raisins were moving away from you.

When scientists began to think about what it would mean if space were expanding outward like raisin-bread dough, they started to come up with the idea that our doughball Universe must have been a lot smaller at one time—and, in fact, that the farther back you went in time, the smaller it was. This is how they came up with the Big Bang idea. The theory goes like this: At the beginning of time (that's what we may as well call it, since before there was space there was no time), all of the matter that now makes up the Universe was squashed into one infinitely dense point. At the moment time began, POW! . . . out it all went, in every direction, constantly expanding, as it is expanding still.

It's even become possible to put a date on the Big Bang, since the theory's early believers said there would be a background "noise" of microwaves left over from the explosion, and that noise has been picked up by Earth receivers. It's coming from 18 to 20 billion light-years away—which means that the event would have taken place that many years ago.

The Big Bang seems like a nice simple theory, but, as we said, it raises a few questions. First of all, physicists have a problem with that infinitely dense point at which the Bang began. They call it a *singularity*, and there is no way to describe it in terms of the physical laws we know. It would be everything and nothing, at once.

But most cosmologists—their motto is "You try and come up with a better idea, Smarty"—accept the Big Bang theory. Some of them feel that from here on, the Universe will continue to expand, which means that the Bang will never really stop. Others believe the Universe has enough mass (including that pesky dark matter) to stop expanding eventually, and perhaps begin to collapse back in on itself. If that happened, would the Big Bang happen all over again? Has it been happening over and over, with no beginning and no end? Will a loss of energy eventually slow the whole thing down to a stop?

There are also those who still believe in a steady-state Universe, one that didn't begin with a Bang but has just always been here. It's expanding, say these cosmologists, because new matter is constantly being created.

But no matter how smoothly the Big Bang or any theory of the Universe's past and future turns out to work, the cosmologists—and the rest of us—are left with the hardest ideas for a human mind to wrap itself around. What does "forever" mean, either in time or space? How can time have had a beginning? An end? And if the Universe has only expanded to a certain point, what's beyond that point?

NOTHING? . . .

... What's that?

WHATSA MATTER WITH ANTIMATTER?

You're walking down the street, and you see somebody who looks just like you. Same face, same backward-facing baseball hat, same T-shirt and jeans, same footwear that your parents think costs too much. You figure you've got to meet this person, so you walk up and shake hands.

Immediately, you both explode.

What happened? Simple—you just met the Antimatter You. For years, science fiction writers have had fun with *antimatter*, strange stuff that real scientists say actually exists. To understand the idea behind antimatter, you've got to understand that all of the particles in the Universe have an electrical charge. Electrons, for instance, have a negative charge. But physicists have discovered that every particle is created along with its opposite—an identical particle having the opposite electrical charge. If you're an *electron*, the strange kid that looks like you is a *positron*.

Now here's the problem. When a particle comes in contact with its opposite, they're both immediately destroyed. So why hasn't the whole Universe blown up? Nobody knows for sure. It seems there should have been as much antimatter, right from the start, as there is regular matter. But we never see any, except now and then in *cosmic rays* (showers of particles from space), and these quickly disappear. Maybe all the antimatter separated and went its own way early on, and there are far-off antimatter galaxies, maybe a whole antimatter universe. It's not likely, but who knows?

In the meantime,
don't shake hands with that kid.

Fancy Word Feature
Impress Friends and Family

positron: The antiparticle of the electron.

antiparticle: A subatomic particle having the same mass, average lifetime, spin, and magnitude of electric charge as the particle to which it corresponds, but having the opposite sign of electric charge, among other things.

juglandaceous: Pertaining to walnuts.

dexteraceous: Pertaining to whatever you want it to pertain to.

STARING INTO THE SKIES

Until barely 400 years ago, the only way to look into the sky was by using your own, unaided eyes. That was the way astronomers had worked since primitive times, and they had accomplished quite a lot. Even the great 16th-century Polish astronomer Copernicus, who explained that Earth revolved around the Sun and so demolished the old notion that everything in the skies revolved around Earth, made his observations without a telescope.

In 1608, Dutch inventors first arranged glass lenses in a tube so that when you looked through the tube, distant objects looked larger. Just one year later, the Italian scientist Galileo heard about the telescope and began to build his own models.

Using the new invention, Galileo began to discover more about the Moon and the planets than had been learned in all the centuries before. He saw the mountains and craters of the Moon as real shapes, and not just shadows. He picked out the individual stars within the great cloudy sweep of the Milky Way, and saw that Jupiter had at least four moons. Looking at the planets, Galileo figured from the differences in their brightness that they varied widely in their distances from Earth.

Most important of all, Galileo proved that Copernicus had been right.

It no longer made any sense to think of Earth as the center of the Universe.

Wow!

67

Since Galileo's time, telescopes have improved tremendously. Reflecting telescopes, which use giant mirrors instead of simple lenses, have reached enormous sizes in the 20th century. The 100-inch-diameter-mirror Mount Wilson telescope that Edwin Hubble used was later joined by the 200-foot Mount Palomar telescope, also in California. In the early 1970s, the Russians built a 236-inch giant that could focus on stars of the 25th magnitude. Put another way, that telescope could see the light from a candle burning 15,000 miles away. (We're not sure whether this means that if you held your birthday party on the Moon, the Russians could see your cake.)

But all the telescopes on Earth have one problem: They're on Earth. Looking up into space from Earth's surface means that you have to look through the atmosphere, which even without smokestacks and car exhausts will keep you from seeing distant objects really clearly. That's why the big telescopes were built on top of mountains like Wilson and Palomar. Now, if you could only get a telescope out beyond the atmosphere . . . .

That's what they did with the Hubble Space Telescope, which was carried into Earth orbit on the space shuttle *Discovery* in 1990. The solar-powered Hubble is a lot smaller than Earth telescopes—only 14 by 43 feet. It's operated by a team of 300 people at a NASA ground station, and receives its instructions by radio. On board are cameras and a spectrograph, which send their images back to Earth as radio messages.

Getting the Hubble to work properly was a little tricky at first—space shuttle astronauts had to go up in 1993 and correct a faulty mirror—but ever since, it's been sending wonderful pictures of objects as near as our fellow planets and as far as the outer reaches of the Universe. The Hubble has taken photos of huge gas clouds giving birth to stars, and has looked at stars so distant that some astronomers have said it's like peering almost to the beginnings of time.

BUT WHO WANTS TO JUST LOOK?

For as long as humans have been looking into space, at least a few of them have thought about visiting the faraway worlds they saw twinkling in the night sky. This kind of dreaming picked up speed when people started to realize that the Moon and planets were *places* you actually might be able to go to, and not *gods* who might not appreciate your visit. As the real shape of the Solar System became understood, the ancestors of today's science fiction writers started cooking up stories about trips to other worlds.

If you look back at any of these tales, you can see that the writers—right up to 100 years ago or so—weren't figuring on two things. First, they failed to realize how fast you have to go—just to escape Earth's gravity (a rocket has to reach 10,800 miles per hour). Second, these writers usually had no idea that there isn't any air in space.

In the second century A.D., a Greek writer came up with a story about a sailing ship carried to the Moon by a whirlwind (the trip took a week, and the sailors met Moon people who traveled on three-headed vultures). In the 1500s, an Italian poet had his hero visit the Moon and its cities in a carriage pulled by four red horses. In an Englishman's story written in 1638, trained geese pull a guy in a chair to the Moon at 85 mph. In 1708, a story appeared in which the Moon voyagers sat in a sled-like vehicle that was shot from Earth by a giant spring. Jules Verne, one of the earliest writers of science fiction who at least tried to be believable, had his late-1800s astronauts reach a Moon orbit after their air-sup-plied space capsule was shot out of an enormous cannon. (If he had had them land, they could never have gotten back.)

"Before man reaches the moon, your mail will be delivered within hours from New York to California, to England, to India or to Australia by guided missles . . . . we stand on the threshold of rocket mail."

   — Arthur E. Summerfield
     (United States Postmaster General), 1959

THE ONLY WAY TO GO

By the early 20th century, though, science fiction writers and scientists began to agree on one thing: Forget the cannons and springs. If you want to travel in space, you need a rocket.

It's likely that rockets have been around for a thousand years. They were probably invented by the Chinese, who also invented what was later called gunpowder. Starting with simple firecrackers, the Chinese went on to build weapons that looked like larger versions of today's bottle rockets. These would explode or set things on fire when they reached their targets, and must have been terrifying to enemies used to ordinary arrows.

Rockets eventually spread to Europe, where they were used in war (think of "the rocket's red glare" in "The Star-Spangled Banner," which describes British ships attacking an American fort in 1814) and for fireworks displays. All of these rockets were propelled in much the same way—with gunpowder or similar chemical mixtures. When properly contained and then lit, the burning fuel thrusts the rocket forward toward its target.

Thrust—there's the key word for describing how rockets operate. A rocket is a *reaction device.* It operates on one of the main principles explained by the great physicist, Isaac Newton: *For every action, there is an equal and opposite reaction.* The thrust of the burning fuel goes backward, and the rocket shoots forward. This is also the way a bullet leaves the barrel of a gun, but there's one big difference: The rocket carries its own fuel, but the bullet gets its push from fuel that is left behind.

All of the rockets we've been talking about so far have been *solid-fuel* rockets. The liquid-fuel rocket, which eventually propelled humans into space, was developed by an American named Robert Goddard, who tested his first successful liquid-fuel model in 1926. Goddard used liquid oxygen and gasoline as fuels. Later, rockets would be fueled by liquid oxygen and liquid hydrogen.

In the 1940s, rockets went back to war. German scientists built the V-2 liquid-fueled rocket, which caused great destruction when thousands of the devices were loaded with explosives and fired at London. The genius behind Germany's wartime rockets was Dr. Wernher von Braun, who surrendered to the United States at the war's end and became a key figure in American space efforts.

GETTING DOWN TO BUSINESS

By the 1950s, it had long been clear that rockets would be the only way to explore space. It was also clear that two countries, the United States and the Soviet Union (the country that Russia used to be the main part of, before it broke up into smaller countries in 1991), would be the ones to make the attempt. The two countries were enemies, often almost to the point of war, and were developing huge military rockets to attack each other with if war ever did come (it didn't, which is why we're all here). Since all of this rocket science was going on, it didn't take a rocket scientist to figure out that racing each other into space would be another way for the two nations to compete.

The Russians got into space first. In 1957, they launched a tiny satellite called *Sputnik*, and in 1961 they sent the first man into orbit around the Earth. Americans had been discouraged by *Sputnik*, and they felt even worse when the first U.S. try at sending up a satellite ended with the explosion of the rocket on the ground. By January

1958, though, America was in space. A Jupiter-C rocket had launched its first satellite, *Explorer I*, into orbit.

The first American in space was a navy officer named Alan Shepard, who rode to an altitude of 115 miles before splashing down in the Atlantic Ocean. That was in May of 1961. In February of the following year, U.S. Marine Colonel John Glenn orbited the Earth three times. The race to the Moon was on.

When Alan Shepard was asked what thoughts ran through his mind as he waited for the countdown that would send him into space, he replied, "I just kept looking around me, remembering that everything in that capsule was supplied by the lowest bidder."

A Really, Really Bad Space Joke

The space program made medical history. It was the first time a capsule ever took a man.

THE RIGHT SIZE

"The foolish idea of shooting at the moon is an example of the absurd length to which vicious specialization will carry scientists working in thought-tight compartments . . . . The proposition appears to be basically impossible."

— A. W. Bickerton
(Professor of Physics and Chemistry, Canterbury College), 1936

When the first seven U.S. astronauts were chosen in 1959, they were all white men—and all military officers who had served as test pilots, trying out new aircraft. Since then, things have loosened up a lot. NASA has sent up women astronauts, black and Asian astronauts, and plenty of astronauts who weren't in the military, and who were trained as scientists instead of pilots. What's really neat is that in 1998, 36 years after his initial flight, John Glenn, at the age of 77, went back into space aboard the space shuttle *Discovery*.

The requirements for becoming an astronaut are more fair nowadays. Even to apply for the astronaut program, you have to have a college degree in math, science, or engineering. You have to pass a tough physical exam, and also do well on the kind of tests that tell whether you get along well with other people (nobody wants to be stuck on the space shuttle with a crab or a weirdo, or somebody who jumps up and down and yells a lot when things go wrong). But there's another set of rules that might seem unfair—if you plan to stop growing soon, or plan to *keep* growing for quite a while.

To be a pilot astronaut on the shuttle, you have to be at least 5 feet 4 inches, and shorter than 6 feet 4 inches. Mission specialists have the same top limit, but they can be a little shorter—as small as 5 feet.

Being short does have its problems. During the planning for a 1997 shuttle mission to the Russian Mir space station, an American woman astronaut found out at the last minute that she couldn't go— the Russians didn't have a spacesuit small enough for her. The male American astronaut chosen to go in her place had his own troubles: the Russians had laid in a food supply based on the tiny woman's needs and tastes, and the guy was a lot bigger. He said he didn't mind.

Really, really, highly recommended reading
The Right Stuff by Tom Wolfe

The Right Stuff. It's the quality beyond bravery, beyond courage. Heroes . . . the first Americans in space . . . battling the Russians for control of the heavens . . . putting their lives on the line.

POGO-STICKING ONE'S WAY TO THE MOON

Just a couple of weeks after Alan Shepard's short flight, President John F. Kennedy decided that America would put a man on the Moon by the end of the 1960s. If you think about it, this is like getting a new pogo stick, finding out that you can manage one jump without breaking your neck, and then deciding that by the end of the week you are going to pogo-stick from New York to Chicago. But the amazing thing is, it happened.

"I believe that this nation should commit itself to achieving the goal, before this decade is out, of landing a man on the moon and returning him safely to the earth."

— President John F. Kennedy
May 25, 1961

America's climb to the Moon was like going up the steps of a ladder. First there were the *Project Mercury* flights, which started with Shepard's little pogo-stick hop and included the orbital flights of John Glenn and other astronauts. Next came *Project Gemini*, named after the constellation Gemini, the Twins. They picked this name because in Project Gemini, each orbiting space capsule held two astronauts. Finally came *Project Apollo* — the final step on the way to the Moon.

Project Apollo was designed to test the equipment that would be used for the Moon landing, especially the vehicle that would carry three astronauts out of Earth orbit, across 240,000 miles of space, and into lunar orbit before lowering the Moon lander to the surface. The high point of *Apollo*, before the actual landing, was the Christmastime 1968 orbit of the Moon—the first time humans had flown beyond Earth's orbit. This was when the first pictures of Earth were sent back from near the Moon, and the first time astronauts got to see the dark side of the Moon—the side that never faces Earth—in person.

The great day came on July 16, 1969, when a 363-foot *Saturn 5* rocket thundered off the ground at Cape Kennedy, Florida, with a thrust equal in power to half a million automobiles. After a four-day flight, the astronauts Neil Armstrong, Edwin "Buzz" Aldrin, and Michael Collins reached the Moon.

The way things were set to work, Collins didn't get to land on the Moon. His job was to orbit in the Command Module, while Armstrong and Aldrin piloted *Eagle*, the lunar lander, to the surface. The lunar lander separated from the Command Module, went down to the Moon, and then returned to the Command Module (just like you might use a rowboat to go from your yacht, into shore for dinner, than back out to your yacht).

It was at 4:18 P.M. Cape Kennedy time, Sunday, July 20, 1969, that the lander settled down onto the dusty surface of the Moon. That's when Neil Armstrong radioed back his famous message, **"Houston, Tranquility Base here.** (They were on a part of the Moon called the Sea of Tranquility.) **The Eagle has landed."** About six and a half hours later, Armstrong set foot on the Moon. His first words after touching the surface have become one of history's most famous remarks:

"That's one small step for a man, one giant leap for mankind."

Great Moments in Chewing Gum History

According to the National Aeronautics and Space Administration (NASA), on June 3, 1965, the Gemini IV astronauts chewed gum in outer space. And what, according to NASA, did they do with the gum once its flavor was all gone? — They swallowed it.

81

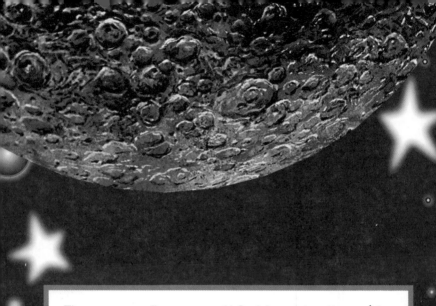

There were five more U.S. Moon landings (the Russians sent only unmanned vehicles), the last one in December 1972. Astronauts stayed on the Moon for as long as 75 hours. They drove around in *lunar rover* cars, little two-seat sports models that ran on batteries. They collected hundreds of pounds of Moon rocks, and brought them back to Earth for study. Alan Shepard, who finally got to go to the Moon ten years after his historic pogo-stick hop, even knocked a golf ball around on the lunar surface using a folding golf club (it's on display at a golf museum in New Jersey).

But since 1972, no one has been back. Moon landings are awfully expensive, and the U.S. turned its attention instead to the re-usable Space Shuttle, to unmanned explorations of other planets, and to plans for a space station.

Feel like writing?
NASA (National Aeronautics
and Space Administration)
300 East Street SW
Washington, DC 20546

Feel like visiting on-line?
www.nasa.gov/

A Really, Really Bad Space Joke

Question: What was the *second* thing
ever spoken on the Moon?

Answer: "Hey, look at that. It's a full
Earth tonight!"

ROCKET STAGES

That monster *Saturn 5* rocket didn't carry the *Apollo* astronauts all the way to the Moon. It would have been way too heavy—and once its fuel was used up, it would have been useless baggage. Rockets used in space missions—even launches into Earth orbit—are built in stages. When each stage has finished firing and has done its work of providing thrust to the main vehicle, it drops back to Earth (this is where the ocean comes in handy) and the next stage fires. The *Apollo* Moon flight depended on three stages of rocket firing—the last stage didn't fall back to Earth, though. It went into orbit around the Sun.

Great Moments in Space!

In 1985, astronauts flew a paper plane in space. It worked just as it does on Earth, only very, very, very slowly. Never mind.

"The acceleration which must result from the use of rockets . . . inevitably would damage the brain beyond repair. The exact rate of acceleration in feet per second that the human brain can survive is not known. It is almost certainly not enough, however, to render flight by rockets possible."

—John P. Lockhart-Mummery, 1936

BEYOND APOLLO

Since the Moon landings, NASA—and the Russians, who haven't landed on the Moon—have held off on sending humans beyond Earth's orbit. The American space program has gone in two basic directions since the *Apollo* days—unstaffed probes of the Solar System, and the Space Shuttle.

The most famous of the radio-controlled unstaffed space vehicles that have been sent on exploration missions are *Voyager I* and *Voyager II*, both of which were launched in 1977. They've taken terrific close-up photos of the outer planets, which have been sent back as radio messages and turned back into pictures on Earth. These deep-space travelers are never coming back. *Voyager II*, in fact, is perhaps more famous for being the first artificial object to leave the Solar System than for doing the work it was designed to do and did so well. Along with *Voyager I*, it's also famous for serving as a sort of message in a bottle. We'll get to this part of the story later, but for now let's just say that the message has to do with a very special phonograph record.

The most exciting recent space probe has been the *Pathfinder* mission to Mars. More than 20 years before, the *Viking* mission had landed a spacecraft on the Red Planet (the nickname for Mars). It sent back pictures, which showed a desert with no canals, cities, or little green men. But *Pathfinder* had better tricks up its sleeve. When it landed on the martian surface on July 4, 1997, it opened up to let a little six-wheeled car—like a dune buggy the size of a toddler's wagon—drive down a ramp and onto the surface.

Back on Earth, a guy was actually controlling the vehicle, which was called *Sojourner*.

He wasn't pointing a remote-control antenna up at Mars and hoping for the best, the way you would if your radio-controlled toy 4 x 4 was six states away and you had really good batteries. Instead he moved an image of *Sojourner* on a computer screen—making the real vehicle move just as he wanted, 119 million miles away, at a speed of one inch every two seconds (not exactly Hot Wheels, but hey, it was on Mars).

Sojourner was equipped with instruments that would let it examine martian rocks and see what they were made of. The big question, for scientists waiting to look at this information back on Earth, was whether any of the rocks *Sojourner* studied have ingredients that could be building blocks of life.

So that's how it finally worked out: After years and years of science fiction stories and movies about martians landing on Earth and saying "Take me to your leader," Earth instead lands on Mars—in the form of this little robot, *Sojourner*—and takes itself to meet martian rocks. Not as exciting as the stories and movies? Maybe not, but it's got something they don't have.

It's real.

85

THE SPACE SHUTTLE

Back in the 1960s, there was a magazine cartoon that showed a repairman handing a scientist a satellite he had just fixed. The caption under the drawing has the repairman saying something like, "That'll be $6,287,356.75. We hadda take it into the shop."

The joke was based on the fact that when a repairman came to fix your TV, it always cost a whole lot more when he "hadda take it into the shop"—so imagine what a satellite repair would cost! Of course, fixing satellites was impossible, because there was no shop to take them to.

When the space shuttle was designed in the 1970s, part of the reason was so we could have that shop. There were a lot of other reasons — the shuttle would be a vehicle that could launch satellites and it would be a good place to perform

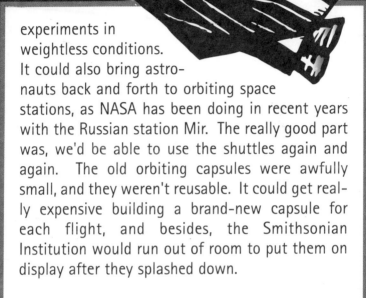

experiments in weightless conditions. It could also bring astronauts back and forth to orbiting space stations, as NASA has been doing in recent years with the Russian station Mir. The really good part was, we'd be able to use the shuttles again and again. The old orbiting capsules were awfully small, and they weren't reusable. It could get really expensive building a brand-new capsule for each flight, and besides, the Smithsonian Institution would run out of room to put them on display after they splashed down.

The space shuttle, which looks more like an airplane than any earlier space vehicle, was designed to be launched on the back of a rocket, orbit on its own (its two solid-fuel boosters and the big central liquid-fuel storage tank for its own engines drop away within nine minutes after launch), and then glide back down into the atmosphere and land on an airstrip.

The first shuttle was launched into orbit in 1981. Five have been built for actual use—an earlier model, named *Enterprise* (after the ship on *Star Trek*) was built just as a test model. There have been more than 80 shuttle missions so far, including the *Challenger* disaster in January of 1986, when a fuel leak caused an explosion that killed all seven astronauts right after takeoff. Other than this tragic and sad disaster, the shuttles have worked well and done what they were supposed to do, carrying weather, communications, spy, and other satellites into orbit, and "taking them into the shop" when they need fixing. Shuttle astronauts can also "space walk" outside their ships to make repairs. This was done in 1993 to fix the Hubble Space Telescope.

Between six and eight astronauts travel on the shuttle at a time. Some are pilots, and some are mission specialists—scientists trained to carry out experiments while in orbit. These experiments have included growing plants (which might come in handy someday on long flights to other planets), manufacturing medicines, and seeing how bees handle flying and making honeycombs in zero gravity (according to NASA, the bees' honeycomb-making attempts were "successful, just as on Earth"). And in case you're wondering, the bees were in special sealed hives. Astronauts don't need loose bees flying around their crew cabin.

Perhaps the most useful of all shuttle activities is one that happens automatically when the ship is orbiting a couple of hundred miles above Earth.

The astronauts get to spend day after day in a state of weightlessness. They eat, sleep, work, and relax without gravity to hold them down. This means they have to learn stuff like how hard to push away from a wall to get where they're going, and how to handle simple objects that always want to go their own way. (Back on Earth they have to re-learn normal behavior, or they'll get into trouble, like the astronaut who dropped his cup to the floor when he forgot that it wouldn't just float in the air). All of this weightlessness training will come in handy when the first planetary explorers head off for months at a time.

Michael Jackson didn't invent the moon walk.

CLOTHES, FOOD, TOILETS (ESSENTIAL!), AND OTHER STUFF

If you look at photos of astronauts working on the space shuttle, you might be surprised to see people wearing shorts and T-shirts, looking as if they're heading out to play tennis. Where are the silver spacesuits?

The answer is that the astronauts don't need them, at least not for their everyday work around the flight deck and crew quarters. The shuttle carries its own air supply. It's also temperature-controlled and kept at the pressure humans need to live. So special pressurized suits with helmets are used only during takeoff and landing, and the real spacesuits, with oxygen and electrical power and a drinking-water tube and air-conditioned underwear, are only used for spacewalks outside the shuttle.

Other things have changed since the early days of space travel. Back in the 1960s, astronauts ate a lot of their food from toothpaste tubes. This was the easiest way to eat on these short missions, without having food float all over the place where there's no gravity. Now, it's a little more like home up there—even if it's a home where all you ever eat is TV dinners.

On the space shuttle, there's enough room for a little kitchen. Crew members take turns getting meals ready, using packages of dried food that they inject water into, or opening cans of moist food. If the "cook" needs to heat anything, there's a microwave oven. The sealed portions are set into pockets in each astronaut's tray, and uncovered little by little to be eaten with regular knives, spoons, and forks. The big differences are that the diners have to use liquid salt and pepper, because separate grains would get loose and float around the cabin (shuttle sneeze time); and also straws for their drinks, because liquid won't pour when it's weightless.

Even in space, what goes in must come out. Yes, there are space toilets. You have to use a seat belt

to sit on one, or you'll float right off. Whatever you left in the toilet will float right out, too, which is why those clever shuttle designers put a vacuum in the bowl. Solid wastes get sucked down into a tank, where they get disinfected and kept until landing. (Yes, there's a NASA guy whose job is to empty the potty tank.)

When an astronaut has to pee, he or she uses something you've probably wished you had in the car. It's a flexible tube that snaps onto a personal attachment, which in turn attaches to the astronaut's even more personal attachment. Then it's vacuum time again. But not a very strong vacuum.

Out on a spacewalk, it's a different story. An old story, one that you would remember if you could remember being one or two years old. Diapers. They're fancy diapers, guaranteed to keep a spacesuit dry, but they're diapers.

You might think that becoming an astronaut is a great way to get to stay dirty for a long time. This used to be true. The early space pioneers kept the same outfits on, but by the time *Skylab* came along in the 1970s, space vehicles were getting big enough to have on-board showers. There are showers on the shuttle today. They're sealed shut while in use, and the water is vacuumed up after an astronaut showers. As for clothes, you can change your underwear every day, and your shirt every three days.

Dirty enough for you?

Now that you've had your shower, it's time for bed. How do shuttle astronauts sleep? First of all, they have to have masks, because the sun comes up five times every eight hours when you're in Earth orbit. As for beds, they have sleeping bags that they strap themselves into to keep from floating around while they snooze. There was one guy, though, who liked to drift about in his sleep. The other astronauts would see him bob past, snoring away. He certainly couldn't say his mattress was lumpy.

ZZZZZZZZZZ

BLAST OFF FOR MARS

As much fun as it is to give little *Sojourner* orders and watch it poke around among the rocks on Mars, working with the robot vehicle only makes space scientists more and more itchy to get up there and do the poking themselves.

Mars, after all, has to be our next stop in the Solar System. Within the next ten years, we will probably have gone through with our plans to build an orbiting space station, and will have sent up a Mars probe that will not only look at the rocks, but also bring back a bunch for us to check out on Earth. But the big prize will still be a staffed landing on the Red Planet.

The big year might be 2011, when the orbits of Earth and Mars bring the two planets close enough so that the trip would take only six months. One of the plans NASA is thinking about would have a team of astronauts stay in a "surface habitat," a station on Mars with an artificial Earth environment, for 500 days. With the habitat as a base, the explorers could head out onto the martian surface. When it was time to head home, they would get on board a

"Mars Ascent Vehicle" that would actually burn fuel made on Mars using hydrogen brought from Earth and gases from Mars's atmosphere. The Ascent Vehicle would take them to their Earth Return Vehicle, which would have been orbiting Mars waiting for them.

Another six months and they're home, arms full with Martian souvenirs. As they come down the ramp to greet the cheering crowds, they hope their luggage hasn't been accidentally sent to Neptune.

But seriously—will this Mars voyage happen? There's no reason why it can't. The trip wouldn't demand any big new inventions that we haven't dreamed up yet—just money. Think about it: If you're a kid reading this book in 1999, you'll get to vote for the president and people in Congress who will make the final decisions about going ahead with the trip.

This means that going to Mars is up to you.

SKYLAB TAKES A DIVE

There's a lot of junk in space—everything from a toothbrush that floated away during a spacewalk, to old satellites that haven't yet re-entered the atmosphere and burned up, to cameras and overshoes and even the lunar rovers that were left on the Moon. Most of this stuff is quickly forgotten.

But for a while back in early 1979, space junk was a hot subject. This was not just because the world was stuck somewhere between disco and Madonna, but because a very large piece of space junk—namely, an orbiting laboratory the size of a trailer truck—was about to fall out of orbit and plow into Earth. Into the ocean, maybe, or your father's garage.

The laboratory was called *Skylab,* and it had been launched by the United States a few years earlier so astronauts could do scientific experiments in space. By 1979, *Skylab* had done its work and was abandoned. It was losing speed and altitude as Earth's gravity pulled it back home. The big question of the summer was, where is this thing going to land?

Skylab got people thinking about "space junk," which includes just about everything that's ever gone up and worn out. Ever since the Russians launched *Sputnik I* in 1957, thousands of satellites have been launched.

The closer they orbit to Earth, the sooner they come down. A satellite entering orbit only 100 miles up, for instance, skids back into the atmosphere in a couple of days. At the other extreme are the satellites that send TV signals. These orbit at an altitude of 22,500 miles, and gravity hardly bothers them. They could keep toddling along, beaming down reruns of *Flipper*, for millions of years.

Old satellites and abandoned space stations aren't the only space junk. Everything from used booster rockets to worn-off space shuttle paint chips, astronauts' kitchen trash, and, ah, some toilet flushings go into orbit and eventually come down. Not *all* the way down, fortunately, since the smaller stuff burns up from friction once it starts penetrating the atmosphere.

Skylab, though, was too big to burn all the way up. On July 11, 1979, it came back to Earth . . . in the Outback of western Australia.

Nobody got hurt, and the souvenir business was fantastic.

REALLY FAR-OUT TRAVEL

So far, we've been talking about travel in and around our own Solar System. But in terms of the Universe and everything in it, this is like looking around in your closet when there are places like the Amazon jungle to be explored. When can we get on with the real *Star Trek* stuff, and break out of our little corner to cross the Milky Way or even cruise to other galaxies? After all, if we're sure there's no other life—at least, intelligent life—here in our Solar System, won't we have to explore other planets that circle other stars if we hope to meet someone worth talking to?

There's This Little Problem . . .

Before you get ready to blast off, think about the distances involved. Remember how far a light-year is? It's how far light travels in one year at its speed of 186,000 miles a second—six trillion miles, to put it in a nice round number. Nothing within the Solar System is anywhere near this far from Earth, but everything out beyond is farther—mostly a *lot* farther.

A Ringside Seat at Gettysburg

With light speed, we have to remind ourselves that we're talking about time as well as distance. If we look up at Alpha Centauri, we're seeing light that left the star 4.3 years ago. When we look through a telescope at a really distant star, we may be seeing it as it appeared when Columbus sailed to the New World (that star would be a little over 500 light-years away), or when Rome was founded (over 2,700 light-years), or even when dinosaurs ruled the Earth (65 million light-years or more).

It gets you to thinking . . .

For three days in early July of 1863, the Union and Confederate armies fought the most important battle of the Civil War at Gettysburg, Pennsylvania. Everybody who took part in the battle—even the survivors—is long dead, so you can't exactly get a first-hand account of what it was like.

Unless, of course, you feel like heading out and watching the battle for yourself.

No, you don't need a bus ticket to Pennsylvania and a time machine. You need a *very* powerful telescope—one that would make the Earth-orbiting Hubble Telescope look like something that came in a cereal box—and a spaceship capable of traveling faster than the speed of light.

Here's the deal. Remember what we said about looking up at the night sky—about how you're not really seeing the stars, but seeing only what they looked like when the light that's hitting your eyes started traveling from them 10 or 100 or maybe even 1,000,000 years ago? That's right—they might not even *be* there anymore.

Well, suppose you could leave Earth at a speed faster than the speed of light—say, fast enough so that you would beat light to a certain point out in space by 136 years. You get to that point, and you whip out your super telescope. You aim it at Gettysburg—no, that's New Jersey; you have to move it a little to the left—and what do you see?

That's right, you see the light—and the image—that left Gettysburg 136 years ago, and is just getting to your telescope right now. There's General Pickett, leading his charge. There's General Chamberlain, defending Little Round Top hill for the North. Look at that—you can even tell he hasn't been shaving recently.

Sound impossible?

Maybe—or maybe we just don't have sharp enough telescopes and fast enough rockets. But if these come along in your lifetime, you might wind up with a ringside seat at Gettysburg. And if you really step on the gas on the way back home, maybe you can catch your great-great-grandfather just before the battle, and tell him when to duck.

Yo, Grandpa. DUCK!

What?

No Time Like the Present — And No Other Time, Period

As for telling your great-great-grandfather when to duck, let's not be silly. You can't talk with your great-great-grandfather any more than you can talk with a photo of him.

Face it, he's history.

Still, time travel is one of the oldest subjects of science fiction. H. G. Wells wrote his wonderful book *The Time Machine* in 1895, and readers have been enjoying its tale of an inventor who can cruise back and forth in time ever since. Fiction writers continue to enjoy working with the time-travel idea, partly because in the more than 100 years since Wells's book appeared, they have come up with ways to use scientific language to make the impossible seem possible. It all has to do with what happens when we apply light speed to the idea of space travel—but do we have any business thinking about going that fast at all?

A UNIVERSAL SPEED LIMIT?

Back in the early 1800s, some people said that Oregon and the other new territories in the far northwestern corner of the U.S. could never become actual states. The reason? Simple—they were so far away, and so hard to get to, that if they did become states, their congressmen would have to spend twelve months a year traveling to Washington, D.C., and back. They would never be able to get anything done in Washington, and never be able to spend any time at home, either. Some people nowadays would say that this would be a great thing to do with politicians.

All this was before the invention of the railroad, which made it possible, by 1869, to travel cross-country in a week. All the West Coast territories became states (California did so even before the railroad ran across the country), and today their senators and representatives can fly home for the weekend.

New inventions, such as the railroad, and later the propeller plane, jet, and rocket, give people the idea that we should never laugh off any ideas of travel, no matter how wacky the ideas are or how great the distances involved. The belief is that somebody will invent something that'll go fast enough for us to be able to go anywhere, even to the stars.

But maybe they won't—and perhaps they can't, at least as far as the stars are concerned. The reason is that covering the distance between our Solar System and any other than the closest stars—unless you want to get there long after you're dead—would involve zipping along at close to the speed of light. Heading for the really far stars would mean hitting the speed of light itself, and breaking it.

But Albert Einstein, the greatest physicist of modern times, said you can't do that.

So who put Einstein in a state police uniform, writing out tickets to light-speeders? Why can't you go as fast as light, or faster?

This isn't the place to explain Einstein's Theory of Relativity, but let's just say that it has to do with how time, space, energy, and matter all relate to one another—and with the fact that energy and matter are really the same thing in different forms. Anyway, one of the things Einstein figured out is that there's a problem with one of the older laws of physics, the one that says that for every action, you get an equal reaction. This is how rockets work. The thrust goes out the back, and the rocket moves forward at the same speed.

What Einstein found out is that the faster you go, the harder you have to push. The difference is too little to matter, or even be noticed, at speeds of just a few thousand miles per hour. But when you start pushing toward light speed—remember, that's 186,000 miles *per second*—it's a different story. It all has to do with mass increasing with speed. Each increase in speed requires more and more energy, since you're speeding up something that has more and more mass. Imagine pedaling your bike up a hill, and the faster you go the heavier the bike gets. No matter what they put in that cereal you ate for breakfast, sooner or later (probably sooner) you're going to get to the point where you just don't have enough energy to pedal any farther—let alone pedal faster.

Now coast downhill, and pay attention. It's the same with the rocket trying to approach light speed. Eventually, you'll be pouring all the energy in the world, maybe in the galaxy, into cranking out more miles per second. And up ahead there's Einstein, holding up his hand and getting ready to write you a ticket. Actually *breaking* light speed, he says, is impossible no matter how much energy you have. This is because at that ultimate speed, you and your spaceship would have reached infinite mass. Try pedaling that *any* faster.

ANY WAY AROUND EINSTEIN?

People don't like limits. Even those of us who don't personally want to travel across the galaxy can get a little upset with being told that we couldn't go fast enough if we *did* want to. And anyone who believes that Earth has been visited by intelligent beings from outside our Solar System doesn't want to hear that it couldn't be true, because those beings are stuck with the same speed limits as us.

So a lot of people—serious scientists, and more than a few crackpots—have come up with theories that offer a way around Einstein's speed limit. Some say that there are in fact particles that travel faster than light, and that if we figure out a way to harness them we have the energy problem licked. Others suggest that maybe we can even transform a starship and its crew *into* such particles for the trip, and then transform them back when they get where they're going. This kind of thinking is behind ideas such as the "Beam Me Up" trick on *Star Trek*, only here it applies to whole space voyages. Good luck—and let's hope you come out in one piece at the end of the trip. (Ever see the movie *The Fly*? It's about a guy who tries to beam himself from one little booth to another, only a fly is in the booth with him, and he comes out half fly.)

This is also where the stuff about time travel comes in—remember your great-great-grandfather at Gettysburg? One of the weird things Einstein figured out about traveling at extremely high speeds is that time would actually go slower for the people doing the traveling than for the people back on Earth (this happens when you're in an airplane, too, but it's far too small a difference to mean anything). "Well," says some of the science fiction crowd, "if time slows down as you approach light speed, does it run backwards if you pass it?"

Or, to put it another way, if you travel away from Earth at speeds faster than light, then hurry home just as fast or faster, do you get back before you left?*

* This sort of thinking often inspires folks in Hollywood; as in the movie *Back to the Future* (and other similar flicks — see page 135 for some of our favorites).

Of course not. Never mind your great-great-grandfather—what would things be like for you if you got back before you were born? Would you disappear? And suppose you gave great-great-grandpa the wrong advice about ducking? With him dead on the battlefield, how are you ever going to get born at all? Think about it. Without him there's no child of him; so there's no great-grandparent, and then no grandparent for you. No grandparent for you means no parents for you, which leaves no you.

So much for time travel. It makes great science fiction, but lousy science.

Other thinkers have wondered if there might be shortcuts through space—*wormholes* or *stargates* that would let you go in one place and come out instantly at another, perhaps millions of light-years away. Maybe black holes, they say, could act this way. (Want to be first to try getting sucked into a black hole?) All this thinking is based on the idea that space doesn't follow the rules of time and distance we know here on Earth, and that it has dimensions we don't quite understand.

Then there are those who figure, fine, we can't travel at light speed or faster, but so what? Space explorers can just be frozen or put into a long sleep (maybe something like *diapouse* — a state of life in which growth and development is temporarily suspended), then take as many hundreds or thousands of years as they need to get wherever they're going. As they near their goal, a wake-up call will bring them back to life. Or, maybe we can just build ships big enough and well equipped enough to carry a whole colony of people who will have babies and die over many generations before they reach another star system.

Time
Machine

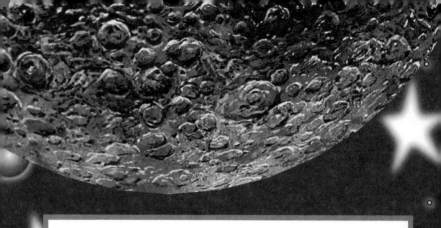

It's easy to see the problems with these ideas. The frozen astronaut plan is nice, but it means that even if the crew re-freezes for the trip back, and doesn't age, everybody they knew on Earth would have been dead for thousands of years by the time they get home. Maybe the whole human species would be extinct. Maybe Earth wouldn't even be habitable anymore.

Might as well stay in space.

As for the traveling colony idea, just imagine umpteen generations of kids, all asking

"Are we there yet? WHEN ARE WE GONNA GET THERE?"

Fancy Word Feature
Impress Friends and Family

diapouse: When growth and development is temporarily stopped.

Instant Creature: A really great Planet Dexter book that's all about diapouse (see page 144).

The Best Things Ever Said About Space

"Exploration is really the essence of the human spirit."

> — Frank Borman, astronaut

"It's a vast, lonely, forbidding expanse of nothing . . . . and it certainly does not appear to be a very inviting place to live or work."

> — Frank Borman, astronaut, during the first orbit of the moon

"How come they picked you to be an astronaut? You got such a great sense of direction?"

> — Jackie Mason

"The moon is essentially gray, no color. It looks like plaster of Paris, like dirty beach sand with lots of footprints in it."

> — James Lovell, astronaut

"I think a future flight should include a poet, a priest, and a philosopher . . . we might get a much better idea of what we saw."

> — Michael Collins, astronaut, on his flight to the moon

"Space—the final frontier . . . These are the voyages of the starship *Enterprise.* Its five-year mission: to explore strange new worlds, to seek out new life and new civilizations, to boldly go where no man has gone before."

> — Gene Roddenberry, creator of *Star Trek*

"Here men from the planet Earth first set foot on the moon, July 1969 A.D. We came in peace for all mankind."

> — plaque on moon marking the first landing

"What is it that makes a man willing to sit up on top of an enormous Roman candle, such as a Redstone, Atlas, Titan, or Saturn rocket, and wait for someone to light the fuse?"

> — Tom Wolfe, *The Right Stuff*

Twinkle, twinkle, little star,
How I wonder what you are!
Up above the world so high,
Like a diamond in the sky.

> —Jane Taylor

Is Anyone Out There?

Of all the questions human beings ask about the Universe, the most exciting doesn't have anything to do with the Big Bang, or dark matter, or how fast the Milky Way is flying toward the Andromeda galaxy. It's the question about whether we are the only intelligent species in the Universe—the only ones to have built a civilization and started asking those kind of questions.

If there are any other civilizations out there, the next questions are: Who are these "aliens" or "extra-terrestrials"? And can we get in touch with each other? These questions have fascinated human beings since at least the time of the ancient Greeks.

Such questions make some people hopeful, other people fearful, and movie directors rich.

What Are the Chances?

There are scientists who believe that the chances of finding intelligent life anywhere out in space are zero—because, they say, we are probably the only life that has ever evolved, anywhere. According to this way of thinking, the recipe for even the simplest forms of early life was so special, and the chances of the ingredients coming together under just the right conditions was so small, that Earth is the only place where life ever happened.

Then there are other scientists who say, nah, life isn't all that hard to cook up. All you need is a planet, or a large enough moon of a planet, and the right atmosphere and ingredients, and the soup starts cooking.

Well, what about planets? We have nine of them in our Solar System, but our space probes have told us that except for Earth, none of them seem like likely candidates for life—unless it's life in a very simple form. But this is just one Solar System. The Milky Way galaxy has 100 billion stars, and astronomers figure there are at least 150 billion other galaxies. Most scientists believe that a vast number of those other stars are the right age and size to have planets orbiting around them. In fact, astronomers have already discovered several stars that have planets.

Here's where the numbers game comes in, and just about anyone can play. Figuring on how many stars there are in the Universe, and how many of them might have planets, and how many of those planets might have the right ingredients and conditions for life, scientists have come up with figures that run from the hundreds of millions up into the billions of billions. The next step, though, is to start figuring how many of those planets with life will develop intelligent life—and how many intelligent civilizations will get to the point where they try to communicate with other worlds.

Well, we know of just one such civilization. It's us, and we've been trying to listen and communicate.

On page 53, there was mention of a constellation known as "Dexter Major (Big Dexter)." You know we were kidding, right? There's no such constellation. Our apologies for any confusion.

TUNING IN TO THE STARS

The listening started for real in 1960, with a scheme called Project Ozma (it was named after a princess in one of *The Wizard of Oz* books). The people involved in Project Ozma figured that the most likely way one civilization would communicate with another—whether on purpose or by accident—was by radio waves, which travel at the speed of light.

Using a 85-foot-wide radio telescope—a receiving device that looks like an enormous satellite dish—the project's scientists focused on two stars for a total of 200 hours. They didn't find any signals that seemed to have come from an intelligent source, but they were pretty sure they knew why their search had failed. It could very well be that they were looking at stars that had no planets with intelligent life. Maybe there was life, but it hadn't developed a civilization that used radio. Or, perhaps they were searching in the wrong area of the radio broadcast spectrum. Radio waves cover a wide range of signal bands, or frequencies. The word *frequency* here refers to the cycle of the radio waves.

Think of it this way: The searchers were looking for a program they weren't sure was on, and they were tuning in to what might have been the wrong channel.

Project Ozma wasn't the end of the search. Since then, scientists working at the SETI Institute (SETI = Search for Extra-Terrestrial Intelligence) have figured out how to "tune in" to many more stars, on as many as 28 million channels at once, using computers that weren't around in Ozma's day. And what are they listening for?

To put it simply, they're listening for basic math exercises. If we were to receive signals in a pattern that couldn't be accidental—say, a series of numbers such as 2, 4, 8, 16, 32, 64, etc.—we could be pretty sure somebody out there was sending them. It could even be something more complicated—as long as it was based on math, which would have to be the universal language of intelligent species everywhere (that means that math homework is also universal).

In fact, we've even sent our own math-based messages. One that went out in 1974 used a mathematical code to tell what our Solar System looks like, what our most important chemical elements are, the size of the Earth's population, and even simple information about DNA—the basic building-block of life on Earth.

But even at the speed of light, that message won't get to the star cluster it was sent to for 24,000 years. Radio signals have the same problem human space travelers would: By the time the round trip is made, there might be nobody around who cared. If SETI picks up an intelligent signal, how do we know the civilization that sent it still exists—or remembers the original broadcast?

Well, at least we can holler out into space and tell whoever's listening that we used to be here. We've even sent out souvenirs.

On the unstaffed spacecraft *Pioneer 10* and *Pioneer 11*, sent out to do research inside our own Solar System during the 1970s, scientists installed plaques with diagrams that tell some basic facts about our home in space, and drawings of male and female human beings. The idea was that since the two *Pioneers* would keep on traveling after they left the Solar System, they might eventually be discovered and studied by another civilization.

Dear Planet Dexter. When people run around and around in circles we say they are crazy. When planets do it we say they are orbiting . . . .

When the *Voyager 1* and *2* spacecrafts were launched in 1977, they carried an even more interesting look at Earth and Earthlings. These two spacecrafts, which were also designed mainly to send back pictures and information about the planets in our Solar System, would also wind up sailing through space for who knows how long and how far. So each one carries a special gold phonograph record (CDs weren't invented yet). On the records are samples of Earth music, including Beethoven, Bach, and rock and roll. There are also natural sounds—thunder, crickets, laughter, humpback whale songs, and even a heartbeat, plus many more.

Unlike normal records, these can carry pictures, too. There are people, flowers, trees, buildings, cars, animals—all of the things we take for granted, but that would look as strange to some civilization on the other side of the galaxy as their stuff would look to us.

We even sent along the equipment for playing the records, with instructions in diagram form. Will it ever get used? Will the records ever be played? No one alive now will ever know — unless one of the Voyagers gets picked up by an extra-terrestrial spacecraft flying near our Solar System, and we get a call asking for more rock and roll.

THEY CAN BE WHATEVER WE WANT THEM TO BE

Since no one has ever seen an alien—or at least proven that they've seen one—we humans have been free to imagine them looking however we want them to look. The same thing goes for how they might act, especially when it comes to dealing with us.

There used to be so many science fiction descriptions of aliens that looked like little green men that some writers now use the term LGMs (short for *little green men*) to mean any extra-terrestrial. Lately, the drawings we have seen of aliens in cheesy supermarket newspapers all have big heads and big, almond-shaped eyes.

But think about it: An intelligent creature from another solar system wouldn't have to be any special color, or have a big head, or even have eyes that look like eyes. And he—or she, if they have hes and shes—certainly wouldn't be a "man."

What *would* an extra-terrestrial look like? That would depend on two things. First, it would have to do with the kind of environment in which it had evolved. What does it breathe? How strong is the gravity on its planet? Look at the differences in environments on Earth, and the different types of human beings they have produced, and you can see that there's no end to the possibilities if we start thinking about conditions on other planets.

Now that we've said all that, let's admit that there are a few things that all intelligent species—at least the kind that can develop advanced technologies—are likely to have. The first is a set of senses that help them get information from their environment. Whether the eyes are almond-shaped or not, seeing really comes in handy (so do touching and hearing). The second is a means of handling things. Dolphins are supposed to be very intelligent, but those flippers are terrible at holding screwdrivers, never mind driving a spaceship.

Of course, if the universal speed limits we talked about are for real, they apply to intelligent species everywhere. In that case, the aliens have no space-ships—at least, not the kind they can come and visit us in. And even if they can blast along at light-speed-plus, would they be able to handle our environment if they did show up? Those little green men, if you ever see any, are probably going to be wearing big green spacesuits, with helmets covering their cute almond-shaped eyes.

EXTRA-TERRESTRIALS WE HAVE DREAMED UP

One thing for sure about our own species—we have very active imaginations. Throughout the years, we've come up with all sorts of ideas about aliens. The funny thing is, we seem to come up with aliens that fit whatever mood we're in at the time.

Back in the 1890s, the English author H. G. Wells wrote the book *The War of the Worlds*, in which an advanced civilization of Martians attacks Earth, intending to wipe out humanity and colonize the place. Humans are primitive compared with the invaders, and nothing can stop them. (We won't tell what finally happens, because it's a good book and you should go read it . . . as soon as you finish this one.) In those days, readers were aware of what powerful European countries had been doing in Africa and parts of Asia—conquering people who could hardly defend themselves against advanced European weapons, and sometimes making slaves of them. It wasn't hard to jump to the next step, and wonder what would happen if Earthlings were the primitive tribe, and Martians— or whoever—were the unstoppable invaders.

In 1938, a radio play based on *The War of the Worlds* caused a panic in the United States. Listeners who tuned in late, missing the announcement that this broadcast was *just a play*, actually thought Martians had landed in New Jersey. Some people even abandoned their homes, and drove off looking for safety. Of course, this was the time when everybody had the jitters about what was happening in Europe, where German dictator Adolf Hitler was threatening to plunge the world into war—which he did, just a year later. So it didn't take much to scare people's socks off.

Aliens that are up to no good have always been popular—the most recent examples were the slimeballs in the movie *Independence Day*. Along the way, though, we've imagined our share of kind, helpful aliens as well. In 1951, a movie called *The Day the Earth Stood Still* featured a robot space-man who came to warn us away from nuclear war. (A few years later, the TV show "The Twilight Zone" did a twist on that, with an episode about an alien who says he has come to help humankind, and even carries a book titled *To Serve Man*. But then somebody discovers it's actually a cookbook . . . .) In the movie *Close Encounters of the Third Kind* the space visitors are peaceful, and in *E.T.* the little rubber character is cuddly, wise, and able to cure just about anything with his magic finger.

What are aliens like?

The question should be, What kind of aliens do we want?

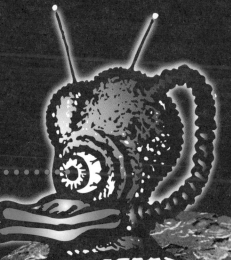

SPACE FLICKS

Sure, at first glance, it may appear that the purpose of space is to provide a place for stars, planets, aliens, astronauts, and flying junk to hang out in. But what space seems to do better than anything else is inspire Hollywood.

Seriously, think about it.

When on Planet Dexter we think of our favorite films, an incredible number of them are inspired by space. Let us know if we've overlooked any of your favorite space flicks (see page 3 for our addresses).

Planet Dexter's Top 20
Space Flicks

(in as-we-thought-of-them order)

1. The *Star Wars* Trilogy: *Star Wars, The Empire Strikes Back, Return of the Jedi* (PG rating)
2. *Men in Black* (PG-13)
3. *Cocoon* (PG-13)
4. *Planet of the Apes* (G)
5. *Apollo 13* (PG)
6. *Space Balls* (PG)
7. *The Right Stuff* (PG)
8. *E.T.* (PG)
9. *Starman* (PG)
10. *Fire in the Sky* (PG-13)
11. *Star Trek: First Contact* (PG-13)
12. *Close Encounters of the Third Kind* (PG)
13. *The Day the Earth Stood Still* (G)
14. *Armageddon* (PG-13)
15. *2001: A Space Odyssey* (G)
16. *Invasion of the Body Snatchers* (both the 1956, black-and-white, not-rated version and the 1978, full-color, PG version)
17. *Stargate* (PG-13)
18. *Independence Day* (PG-13)
19. *Lost in Space* (PG-13)
20. *It Came from Outer Space* (not rated)

"I was kind of lonely at the time. My girlfriend was back in Los Angeles. I remember saying to myself, 'what I really need is a friend I can talk to — somebody who can give me all the answers.'"

— Steven Spielberg, on the birth of E.T.

"Steven's room was a mess. Once his lizard got out of its cage, and we found it, living, three years later. He had a parakeet he refused to keep in a cage. Every week, I would stick my head in his room, grab his dirty laundry and slam the door. If I had known better, I would have taken him to a psychiatrist—and there never would have been an E.T."

— Leah Spielberg (Steven's Mom)

The sound of E. T. walking was made by someone squishing her hands in Jello.

Thanks, Max.

Another Dexterdrome!
Yo! Bozo boy!

An Almost in Space Feature

WARNING: Don't ever try this!

Here's an idea. In 1982, a guy from North Hollywood named Larry Walters strapped 45 helium-filled weather balloons to his aluminum lawn chair and took off from his girlfriend's backyard in San Pedro, California.

Larry thought he'd simply float several hundred miles across the desert. Oops! Instead of softly floating, Larry shot straight up out of the backyard. He went so high—16,000 feet—so quickly that he turned numb and his breathing began to fail. Larry reached for the gun he had brought along (just to be safe, we guess). He started shooting the balloons; he began to descend; he—darn!—dropped the gun after taking out only seven balloons.

Larry floated downwind for miles until a Delta pilot, busily flying over Long Beach, looked out his cockpit window and saw Larry in his lawn chair. A police helicopter was alerted and it followed Larry until he slammed into a driveway, a bit shaken but otherwise OK.

Walters' flight broke the world altitude record for clustered balloon flight, but because it was unlicensed and unsanctioned, the record is not official. **Darn!**

HAVE THEY ALREADY BEEN HERE?

It isn't unusual nowadays to find people who aren't wondering about whether there are advanced societies somewhere out in space, because they believe that Earth has already been visited by extra-terrestrials.

People have seen strange lights in the sky since Biblical times. They saw them in the Middle Ages, and they see them today. Of course, if you didn't know what the Moon was, you'd think it was a strange light, too.

We all know what the Moon is. But what about other lights? Do we all know what every bright planet, unusual aircraft, silvery weather balloon, or other moving or glowing object in the sky is? An *Unidentified Flying Object*—a UFO—is just that: An object that is flying (or at least up in the air) that we haven't identified yet.

But people love mysteries, and they love to jump to the strangest explanation first, rather than last. This is just the opposite of what science teaches us, namely that you're supposed to go with the simplest explanation first. If that turns out to be false, go to the next simplest. And so on.

Well, that's no fun. And because it's always more interesting to have a mystery—or to think about little green men—we have thousands upon thousands of UFO reports.

Remember how the radio broadcast of *The War of the Worlds* came along just before World War II, when people were getting nervous about real armies and real invasions? The big wave of excitement about UFOs came not long after that war was over, when people were hearing more and more every day about jet planes, nuclear bombs, and the coming "space race" between the Soviet Union and the United States.

Another Dexterdrome!

Neil, an alien.

The term *flying saucers* was first used in 1947, when the pilot of a private plane reported seeing things that looked like "pie plates" in the sky. In that same year, the Roswell legend got its start. People near Roswell, New Mexico, claim they saw a strange object crash, and some of them found unearthly-looking debris on the ground. The government later said it was the remains of a weather balloon. But the rumor persisted that military investigators had gotten to the crash site first, and quickly carried away the bodies of dead alien astronauts. Quite a few people think those alien bodies are still preserved at a secret base in southern Nevada—and that the government just might be hiding an extra-terrestrial ship as well.

From the late 1940s to the late 1960s, more than 12,000 reports of UFOs were collected and investigated by the government—and only 700 or so could not be accounted for by explanations as simple as weather balloons, military aircraft, optical illusions caused by weather conditions, or a really bright showing of the planet Venus.

How do we account for those 700 sightings, and any others since, that haven't been explained? We can't. But as one science writer wrote, the police don't solve all the crimes they investigate, either—but that doesn't mean the crimes they don't solve were committed by space aliens.

What about the pictures some people claim to have taken of UFOs? You've probably seen one or two of these, maybe in one of the weekly newspapers that carry stories like "President Kennedy Alive; Meets with Elvis to Plan Comeback Tour." The funny thing is, the pictures are always either very blurry and could be just about anything, or they look as if they could have been easily faked. Remember, the one famous photo of the Loch Ness Monster—which wasn't that clear to begin with—finally turned out to be phony. Somebody admitted to setting up the shot, and fooling the world for decades.

You can take your own "UFO" photo. All you need is a Frisbee, a tuna fish can, some duct tape (real space aliens use duct tape, so you should too), some silver paint, and a camera with an adjustable focus. Tape the can onto the top of the Frisbee, and paint the whole thing silver. Next, adjust the camera so it's out of focus for the distance you'll be shooting at. Pick a time of day when the light isn't very good—or else plan the shot so the camera will be pointing toward the sun, instead of away from it the way you're supposed to take pictures. Have someone toss the Frisbee so it will be at least 50 feet away when you click the shutter, and make sure the background is something a few hundred feet in the distance—a good idea would be to have the Frisbee over house-tops or trees, so the picture will make it look like it's a lot bigger but farther away.

Now, shoot away. The combination of bad light and uncertain distance will make it look like there's something strange in the sky. Of course—a flying saucer!

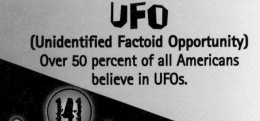

UFO
(Unidentified Factoid Opportunity)
Over 50 percent of all Americans believe in UFOs.

And then there are the stories of alien abductions—of people who claim they were taken up into spaceships and interviewed or examined by extra-terrestrials. These tales leave us with no choice but to believe whatever we want to believe. No one, so far, can prove they were carried off in an alien craft, and yet no one can prove that this never happened. But if someone comes along with a story like this, the first thing a careful investigator has to do is decide if there might be any other reason the "captive" might be missing a couple of hours out of his or her life.

The most far-fetched alien stories of all have to do with visitors from space not only interfering with individual Earthlings, but with the development of our civilization. Here are just two examples: For a long time, people have been amazed that the ancient Egyptians were able to build the pyramids without the machinery that we would use for moving and lifting giant blocks of stone. Another puzzle is the collection of enormous symbols and designs laid out on the floor of the Peruvian desert near the town of Nazca.

In the 1970s, a few writers started to offer the idea that "ancient astronauts"—aliens who were cruising around, and had nothing better to do than get involved with somebody else's building projects—had helped the Egyptians build the pyramids, and planned out the "Nazca lines" as a system of flying-saucer landing strips. (After all, the patterns—mostly crude animal designs— can be understood only from above, and the Peruvians of a thousand years ago didn't have airplanes.)

The fact is, though, the Egyptians did have the technology to build the pyramids. It consisted of a really large number of slaves. As for Nazca, the lines really are mysterious. The New York City subway system would look mysterious, too, if you were looking at it a thousand years after it was built and didn't know what it was for. Who could have created such a thing?

Must have been space aliens.

We are clever enough to have built and learned quite a bit over the past few thousand years. Over the past century alone, we have made enormous progress in figuring out how vast and how old the Universe is, and in understanding the physical laws that it moves by. We have a lot left to learn, including the secret of keeping ourselves and our Earth in good working order. If, over the course of all the centuries yet to come, we meet up with other intelligent creatures who can help us—or whom we can help—that would be wonderful.

But it looks like we've made it this far on our own.

UFO

(Unidentified Factoid Opportunity)

Jimmy Carter was the first President of the United States to see an UFO. In 1969, Carter and a few buddies saw a sometimes bluish, sometimes reddish saucer moving across the evening sky. "It seemed to move toward us from a distance," Carter noted, "then it stopped and moved partially away. It returned and departed. It came close . . . maybe three hundred to one thousand yards away . . . moved away, came close, and then moved away . . . I don't laugh at people anymore when they say they've seen UFOs."

MORE SPACE* BOOKS FROM PLANET DEXTER!

(* Well, sort of. The more of these books you have, the more space they need. And the more of these books a parent buys for you, the more space there will be in that parent's wallet. Also, on the pages of all these books are places where there is no illustration, photography, or words; in the book business, we refer to those empty places as "space.")

"Whaddaya Doin' in There?"
A Bathroom Companion (for Kids!)
by The Editors of Planet Dexter

All really smart kids read in the bathroom. But what to read? Answer: This book. "Whaddaya Doin' in There?" offers humor, bathroom lore, ghost stores, weird laws, lotsa trivia, you name it. It's the perfect tome for your toilet time.

Grossology
The Science of Really Gross Things!
by Sylvia Branzei and illustrated by Jack Keely

Yup, it's slimy, oozy, stinky, smelly stuff explained. **Grossology** features the gag-rageous science behind the body's most disgusting functions: burps, vomit, scabs, ear wax, you name it. Who could ask for anything more?

Instant Creature!
The Swimming Critters from Way Back Then
By The Editors of Planet Deter

Includes creature eggs and official pet snack food. These creatures are alive! Just add water and within days you'll have creatures swimming the backstroke, practicing somersaults, and reproducing. "When I went to buy **Instant Creature**, and the bookstore didn't have it, I refused to buy Nickelodeon's sea monkeys book." — Hune, age 11